FAO ANIMAL PRODUCTION AND HEALTH

guidelines

THE FEED ANALYSIS LABORATORY: ESTABLISHMENT AND QUALITY CONTROL

Setting up a feed analysis laboratory, and implementing a quality assurance system compliant with ISO/IEC 17025:2005

Authors
L.H. de Jonge
Animal Nutrition Group
Wageningen University
The Netherlands

F.S. Jackson
Manager, Nutrition Laboratory
Massey University
New Zealand

Editor
Harinder P.S. Makkar

FOOD AND AGRICULTURE ORGANIZATION OF THE UNITED NATIONS
Rome, 2013

Recommended Citation

de Jonge, L.H. & Jackson, F.S. 2013. *The feed analysis laboratory: Establishment and quality control. Setting up a feed analysis laboratory, and implementing a quality assurance system compliant with ISO/IEC 17025:2005.* H.P.S. Makkar, ed. Animal Production and Health Guidelines No. 15. Rome, FAO.

The designations employed and the presentation of material in this information product do not imply the expression of any opinion whatsoever on the part of the Food and Agriculture Organization of the United Nations (FAO) concerning the legal or development status of any country, territory, city or area or of its authorities, or concerning the delimitation of its frontiers or boundaries. The mention of specific companies or products of manufacturers, whether or not these have been patented, does not imply that these have been endorsed or recommended by FAO in preference to others of a similar nature that are not mentioned.

The views expressed in this information product are those of the author(s) and do not necessarily reflect the views or policies of FAO.

ISBN 978-92-5-108071-9 (print)
E-ISBN 978-92-5-108072-6 (PDF)

© FAO, 2013

FAO encourages the use, reproduction and dissemination of material in this information product. Except where otherwise indicated, material may be copied, downloaded and printed for private study, research and teaching purposes, or for use in non-commercial products or services, provided that appropriate acknowledgement of FAO as the source and copyright holder is given and that FAO's endorsement of users' views, products or services is not implied in any way.

All requests for translation and adaptation rights, and for resale and other commercial use rights should be made via www.fao.org/contact-us/licence-request or addressed to copyright@fao.org.

FAO information products are available on the FAO website (www.fao.org/publications) and can be purchased through publications-sales@fao.org.

Contents

Contents   iii
Foreword   vii
Abbreviations   viii
Acknowledgements   ix

CHAPTER 1
Introduction   1

1.1 Background   1

1.2 Aim   1

1.3 A road map of the document   2

CHAPTER 2
Development of a business plan   3

2.1 Introduction   3

2.2 Approach for development of a business plan   3
    2.2.1 Type of laboratory to be developed   3
    2.2.2 Market analysis for potential customers and their needs   4
    2.2.3 Types of analyses   6
    2.2.4 Market analysis for available laboratory services   6
    2.2.5 Evaluation and decision making   7

CHAPTER 3
Setting up and running the laboratory   11

3.1 Introduction   11

3.2 Physical realization of the laboratory   11
    3.2.1 The analytical process   12
    3.2.2 Methods to be made operational   13
    3.2.3 Building and facilities   18
    3.2.4 Equipment   24
    3.2.5 Organizational structure and responsibilities of personnel   26

3.3 Realization of the laboratory – Procedures   27

3.4 Continuity and improvement of the laboratory   28

CHAPTER 4
Implementation of a Quality Management System and the road to accreditation — 31

4.1 Introduction — 31

4.2 Basic principles of quality — 31
    4.2.1 Technical level — 32
    4.2.2 Organization level — 37
    4.2.3 Commercial level — 40

4.3 Reading and interpretation of ISO/IEC 17025:2005. — 41

4.4 A road map for building a high quality system — 50

4.5 First situation: Routine stand-alone feed analysis laboratory — 50
    4.5.1 Introduction — 50
    4.5.2 Initial phase — 51
    4.5.3 First year — 52
    4.5.4 Second year — 52
    4.5.5 Third year — 54
    4.5.6 Fourth year — 56

4.6 Second situation: Routine laboratory connected to a feed manufacturer — 58

4.7 Third situation: Laboratory as part of a research organization — 59

4.8 Fourth situation: Government or reference laboratories — 60

Sources used — 63

APPENDIX A
Ensuring quality analytical performance — 65

APPENDIX B
First line of quality control and the use of Shewhart charts — 73

APPENDIX C
Validation requirements — 77

APPENDIX D
Calculation of uncertainty of measurement — 79

APPENDIX E
An example of technical records for a determination — 81

APPENDIX F
An example of a maintenance and calibration document   83

APPENDIX G
An example of a training record   85

APPENDIX H
Procedure for traceability of volumetric calibration   87

APPENDIX I
Trend analysis   89

Foreword

Feed has a fundamental influence on productivity, health and welfare of the animal. Feed quality influences animal product quality and safety, and the environment. To achieve balance among these parameters, the animal's nutritional requirements must be properly met.

Confidence in the nutritional information on any feed or feed ingredient provided by suppliers is critical for buyers because it provides a guarantee of feed quality. Current reports from many countries suggest that manufacturers and buyers do not always have confidence in the data provided from non-accredited laboratories, which can negatively affect market prices and international trade. It is therefore important that laboratories work towards adopting a Quality Assurance System for all of their routine feed analyses. This has been detailed in two FAO *Animal Production and Health Manuals:* No. 14, *Quality Assurance for Animal Feed Analysis Laboratories,* and No. 16, *Quality Assurance for Microbiology in Feed Analysis Laboratories.*

Not only must the methods used be of an internationally recognized standard, but all steps in the process, from the initial sample submission through to the final report preparation, must be traceable. An internationally accredited laboratory gives producers and buyers of feed a great deal of confidence in the data they receive. This provides wider market possibilities for feed manufacturers. Also, the right nutritional information about feed ingredients and feeds will enable preparation of balanced diets that meet the nutritional requirements to match the physiological stage of animals and to satisfy the farmer's husbandry objectives.

This document presents a step-by-step process to guide the laboratory management team through the various stages, from planning the feed analysis laboratory building and layout, to hiring suitable staff and choosing which methods to set up, with appropriate equipment requirements. A detailed guideline for initiating a Quality Management System starts with validation of methods, personnel and training; addresses systematic equipment maintenance, calibration, proficiency testing and quality control procedures; and final reporting and auditing, all culminating in a final accreditation inspection within an estimated four-year time frame.

The authors have extensive laboratory experience as well as personal experience with successfully bringing non-accredited laboratories up to an internationally recognized accreditation standard. The content of the document has been peer reviewed by a large number of experts and their suggestions incorporated. The guidelines presented will assist governments and feed manufacturers, as well as a range of institutions, including research and education, to work towards establishment of a feed analysis laboratory – whether as an integral unit or as an independent commercial laboratory – with internationally recognized accreditation.

Berhe G. Tekola
Director
Animal Production and Health Division

Abbreviations

A.U.	Absorption Unit
AAS	Atomic Absorption Spectroscopy
ANOVA	Analysis of variance
CRM	Certified Reference Material
CUSUM	Cumulative sum
FAO	Food and Agriculture Organization of the United Nations
FAO/IAEA	FAO/IAEA Agriculture and Biotechnology Laboratory
GC	Gas chromatography
GC-FID	Gas Chromatography-Flame Ionization Detector
GC-MS	Gas Chromatography-Mass Spectrometry
HPLC	High Performance Liquid Chromatography
IAEA	International Atomic Energy Agency
ICP	Inductively Coupled Plasma [Analysis]
ICP-AES	Inductively Coupled Plasma–Atomic Emission Spectroscopy
ILAC	International Laboratory Accreditation Cooperation
ISO	International Organization for Standardization
LC-MS	Liquid Chromatograph–Mass Spectrometer
LIMS	Laboratory Information Management System
LOD	Limit of Detection
LOQ	Limit of Quantification
MS	Mass Spectrometer
MS-MS	Sequential mass spectrometry
MU	Measurement of uncertainty
NIR	Near-Infrared Spectrometry
PR	Public Relations
PSG	Project Steering Group
QA	Quality Assurance
QMS	Quality Management System
R&D	Research and development
ROI	Return on Investment
SD	Standard deviation
SOP	Standard operating procedure
UV	Ultraviolet

Acknowledgements

We thank all peer reviewers, listed below, for taking time to critically read this manual and for their comments and suggestions that led to its improvement.

Copy editing and preparation for printing was by Thorgeir Lawrence and layout was coordinated by Claudia Ciarlantini, which are thankfully acknowledged.

Jim Balthrop	Quality Assurance Manager, Office of the Texas State Chemist, P.O. Box 3160, College Station, Texas 77841, United States of America
Richard A. Cowie	Senior Quality Assurance Manager, SRUC, Ferguson Building, Craibstone Estate, Aberdeen AB21 9YA, Scotland, United Kingdom
Johan DeBoever	Senior Researcher Feed Evaluation, Institute for Agricultural and Fisheries Research, Animal Sciences Unit, Scheldeweg 68, 9090 Melle, Belgium
E. Fallou Guèye	Animal Production Expert, Animal Production and Health Division, FAO, Viale delle Terme di Caracalla, 00153 Rome, Italy
Gustavo Jaurena	Professor, Animal Nutrition, Centre of Research and Services in Animal Nutrition, Facultad de Agronomía, Univ. de Buenos Aires, Av. San Martín 4453 - C1417 DSQ, Ciudad Autónoma de Buenos Aires, Argentina
Alicia Nájera Molina	International Quality Assurance Manager, Masterlab bv, The Netherlands
E. Odongo	Animal Nutritionist, Animal Production and Health Section, Joint FAO/IAEA Division, IAEA, Vienna, Austria,
Alfred Thalmann	Elbinger Str. 10c, D 76139 Karlsruhe, Germany (Formerly: Staatliche Landwirtschaftliche Untersuchungs- und Forschungsanstalt Augustenberg; later Landwirtschaftliches Technologiezentrum Augustenberg)
Anja Töpper	Head, Animal Feed Analysis and Microbiology Unit, Landwirtschaftliches Technologiezentrum Augustenberg, Neßlerstraße 23-31, D 76227 Karlsruhe, Germany

Chapter 1
Introduction

1.1 BACKGROUND
The importance of livestock production in developing countries is increasing due to growing demands for animal products, combined with demands for more sustainable agriculture. Globalization and increased demand for animal products also offer export opportunities, leading to improved welfare for people from the exporting countries. Key elements for success, however, are high productivity and low prices, coupled with high quality and safety of animal products. Chemical analyses of diets and animal products play an essential role in achieving optimal production and guaranteeing safety for consumers. In developed countries, it is common practice to undertake most of the analyses in accredited laboratories. The availability of this kind of service in developing countries is constrained due to fewer laboratories and a lack of infrastructure. This can lead to a lack of reliable data, which can make animal agriculture in developing countries much less competitively priced when compared with that in developed countries.

The situation in developing countries can be improved by increasing the number of laboratories and improving the quality of the available analytical services. The provision of relevant information and training can be used as a tool to improve the quality of the analytical services in these laboratories. For this purpose, an international group of laboratory experts participated in an FAO working group that led to the publication of two manuals describing quality assurance, safety issues and analytical methods: *Quality assurance for animal feed analysis laboratories* (FAO, 2011), and *Quality assurance for microbiology in feed analysis laboratories* (FAO, 2013). This group was also the starting point for an FAO network of international experts, aimed at improving analytical capability in developing countries.

Increasing the number of laboratories, however, is a challenging task. The initiative to start a laboratory lies with local stakeholders and is based on an investment decision. The high investment and the technical complexity can result in hesitation by stakeholders because of a lack of expertise.

1.2 AIM
The aim of this present document is to present guidelines for starting and running an animal feed analysis laboratory, including the implementation of quality assurance (QA) systems compliant with an international standard. To achieve this goal, the relevant information will be described and illustrated by giving examples wherever appropriate, which should lead to a better understanding by semi-technical persons and decision-makers.

1.3 A ROAD MAP OF THE DOCUMENT

The document is divided into three parts.

The first part (Chapter 2) describes the initial phase of building an animal feed analysis laboratory, a phase especially important for decision-makers and business developers. Some critical and crucial decisions have to be made during this phase, with such central questions as: *What should the laboratory do?* and *What are the chances of success in terms of being an economically viable unit?* To answer these questions, stakeholders should first undertake a market analysis and identify any potential customers, their particular requirements, and other business opportunities to make the project viable. This information is necessary to make a balanced judgement of the potential benefits or profits with respect to the investment required. This document addresses the issue from the point of view of different types of laboratory. The investments needed for facilities, equipment, consumables and labour (time and skill) will be given for the different types of analyses, including a global price indication. Using this information, stakeholders should be in a better position to calculate Return on Investment (ROI).

The second part (Chapter 3) of the document deals with the physical creation and running of the laboratory, and is especially important for financers and the laboratory staff. The first section deals, among others, with the physical creation of the laboratory, such as securing the land; building the laboratory, including the necessary facilities; employing suitable laboratory personnel; purchasing appropriate equipment; and organizing the laboratory (placement of various items of equipment and their safe operation). Special attention is given to pertinent legislation, and health and safety aspects. The second section focuses on the organization of the primary process, from receiving sample material to sending analysis reports and invoices, and management of the laboratory. Crucial elements, such as storage, planning, traceability and confidentiality, are highlighted, and examples of the related procedures described. Management aspects are separated into internal and external. Internal aspects focus on optimization, and increasing efficiency and quality of the laboratory processes. This also includes human resource management. External aspects focus on improving the market position of the laboratory, including enhancing contact with customers, identifying new customers, better positioning of the laboratory within the marketplace, establishing contacts with national and international networks on laboratory analyses, participation in conferences that address feed and food analysis issues, and exhibitions of laboratory equipment. Both aspects are important to ensure the successful running of the organization.

The third and final part of this document (Chapter 4) describes the implementation of a QA system, which is especially important for laboratory staff and the QA manager. The quality of the analytical results produced should be guaranteed to gain the confidence of customers. The implementation of a complete quality control system can take several years of experience and refinement before it can be accredited. In this document, activities and their time schedule are described for the implementation of quality principles with the aim of achieving laboratory accreditation according to ISO/IEC 17025:2005 *General requirements for the competence of testing and calibration laboratories,* which is an internationally recognized standard for quality systems within testing laboratories. Additional information is also given about implementation of these principles in daily practice.

Chapter 2
Development of a business plan

2.1 INTRODUCTION
Preparation of a business plan is essential for the creation of a new laboratory. A business plan describes aims of the laboratory and a plan to realize them. It also assesses challenges in achieving the aims and making a profit, as well as identifying potential opportunities for the laboratory. The plan can be seen as a road map for the future laboratory to achieve its final goal. The physical creation of a laboratory needs a relatively large investment, and therefore a good business plan can also help to increase the confidence of potential investors. The realization of a business plan involves a range of experts, including marketing and technical specialists, to provide a realistic plan for the new laboratory on which the necessary investments can be based.

The absence of a business plan can lead to incorrect decisions regarding investments, or to missed opportunities, which could lead to a non-profitable outcome.

Two different areas of expertise are necessary:
- knowledge to assess the market situation and to identify opportunities and challenges; and
- laboratory knowledge to evaluate types and requirements of potential analyses and the estimation of laboratory costs.

A group to develop a business plan should be created to include personnel with expertise in both the above two areas. They should regularly interact while developing the plan.

A suggested approach to realize a business plan is described in the next section. This approach should not be seen as definitive, but rather as a possible tool to create a business plan. The example approach is limited to chemical analyses.

2.2 APPROACH FOR DEVELOPMENT OF A BUSINESS PLAN
The key elements required to prepare a business plan are given in Figure 2.1.

2.2.1 Type of laboratory to be developed
The characterization of the laboratory to be created is the first important step. From a commercial point of view, a distinction should be made between a laboratory as a stand-alone unit, and one embedded in a larger organization. A laboratory could be one of five types:
- **Stand-alone commercial laboratories.** These laboratories have an independent judicial position, working on a commercial basis with the aim to make a profit. In practice this means that they largely analyse samples sent from commercial clients.
- **Laboratories integrated with feed producing units.** These laboratories are part of a larger organization, but operate as a separate unit within the total chain of quality control of animal diets and pre-mixes produced. In practice they are used for

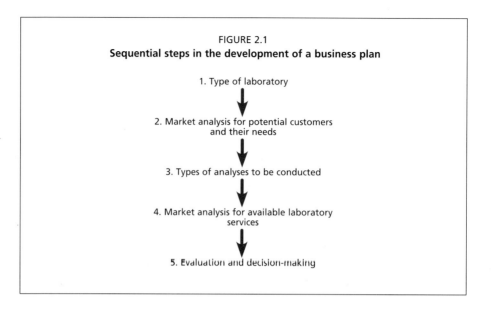

FIGURE 2.1
Sequential steps in the development of a business plan

1. Type of laboratory
2. Market analysis for potential customers and their needs
3. Types of analyses to be conducted
4. Market analysis for available laboratory services
5. Evaluation and decision-making

the control of feed ingredients and animal diet specifications. Sometimes they also provide services to others outside their own organization.
- **Laboratories integrated with research organizations.** These laboratories are used to analyse samples from experiments performed within a research organization. The analytical work conducted in these laboratories is a mixture of standard analyses and more specific research analyses in a wide range of matrices. These laboratories can operate as separate units or be fully integrated within a research division.
- **Laboratories integrated with educational organizations.** The primary function of these laboratories is to provide training opportunities for students. Students may also analyse their own research samples.
- **Government or reference laboratories.** The function of these laboratories is to be a reference for other laboratories and thereby assist in maintaining and improving the quality of analytical work conducted in the individual laboratories within a country. They are also used by the government for regulatory analyses regarding feed and food safety.

In some situations the laboratories connected to research and educational organizations also analyse samples from commercial clients.

2.2.2 Market analysis for potential customers and their needs

The next step is to identify and quantify potential customers and their needs. These could be producers or users of feed ingredients, such as feed industries, researchers, non-governmental and governmental organizations. The number and type of clients may vary between different locations and reflect the level and sophistication of agriculture and livestock activities in the region. Customers are critical for the revenue of the laboratory and therefore its commercial existence. Good market analysis is necessary to assess the opportunities for a new laboratory to be economically viable. This analysis should take into account the maximum acceptable duration for transport of samples to the laboratory (ideally no more than one or two days). It should be noted that differences in regulations between countries may present difficulties in sending samples across international borders.

The market analysis should be performed using the steps presented below, starting with an initial analysis based on the current use of analytical services. This should focus on:
- Number and type of customers.
- Type of analyses and the amount spent on analytical services.
- Organization of the analytical services: in-house versus outsourced analysis of samples.

Customers can be divided into categories relating to the amount they are likely to spend per year on analytical requirements, such as small-scale farmers; small- and medium-scale feed manufacturing units; and large feed manufacturing mills. Analytical services should be divided into different types of analyses. An obvious division is to separate the analytical services into proximate analyses (i.e. the classic animal feed analyses) and advanced analyses that use sophisticated instruments (examples being minerals, amino acids and contaminants). The amount spent on each type of analysis can be roughly calculated by identifying the number of samples and the prevailing market prices.

A division between analytical services performed within the organization and those outsourced to independent commercial laboratories should be made. Some companies may have the capability to perform routine analyses in-house, such as nitrogen analysis, which is likely to continue even if a new laboratory offers the same service. The ratio between both types of services strongly affects the opportunities of the new laboratory.

The second step should assess additional and potential opportunities in the market that are likely to utilize the services of the laboratory being established. Attracting new clients should be a primary goal, and this could be achieved by offering a more comprehensive testing facility, faster turnaround time or shorter travel distance than is currently available in the area, or assisting with interpretation of the results and providing recommendations about ration formulation.

A critical evaluation should be undertaken to establish why these potential clients would use analytical services in the future, but are not currently using the services. Having the ability to perform some unique analyses offers clear opportunities for the laboratory and could strengthen its market position. Its realization, however, will have an impact on investments needed (see Section 2.2.3).

The third step is to investigate and predict future market developments. Some important points that can be addressed are:
- growth of the animal production sector;
- pressure to produce animal products efficiently and sustainably;
- national and international legislation for feed and product safety; and
- volume of feed or feed ingredient export.

These issues will require the generation of new information and data, and therefore the need for analytical services. As part of the market analysis it is important to seek advice from local councils as well as national regulators to understand market trends and possible changes to legislation. These should be taken into account at the planning stage. This market analysis should be performed for each type of laboratory.

2.2.3 Types of analyses

The types of analyses conducted by the laboratory affect both its market position in terms of attracting potential customers, and the investment needed. The types of analyses can be divided into five types:

    Type 1. Proximate analyses
    Type 2. Macro-minerals
    Type 3. Micro-minerals at trace level
    Type 4. Chromatographic analyses (e.g. amino acids, fatty acids)
    Type 5. Chromatographic analyses at trace levels (contaminants such as aflatoxins, pesticides and pesticide residues, antibiotics, etc.).

The types of analyses will determine the investment needed. Proximate analyses are used for feed characterization for general nutritional parameters, and the capacity to perform these analyses should be seen as the minimum requirement for every laboratory. Other types of analysis are more specialized and need specific equipment and facilities. For minerals and chromatographic analyses, it is important to make a distinction based on the required detection limit of the samples to be analysed. Determination of trace levels are mostly performed to establish the presence or absence of a contaminant that could affect public health, which governments try to protect by legislation (e.g. aflatoxins, pesticides, pesticide residues and antibiotics). Consequently, these determinations not only require highly skilled personnel and sensitive and expensive equipment, but also demand a higher level of purity of chemicals used (including water) and clean work conditions to avoid contamination.

Types of laboratories can be tentatively categorized as:
- Basic nutrition laboratory performing only proximate analyses (Type 1 analyses).
- Laboratory conducting analysis of nutrients; performing proximate, mineral and chromatographic analyses (Types 1 to 4 analyses).
- Laboratory conducting analysis of nutrients and anti-nutrients (Types 1 to 5 analyses).

All animal nutrition laboratories should be able to perform proximate analyses, with the possibly of extending to analysis of other analytes in the future.

It is also possible for a laboratory to sub-contract some analyses if it is not economically viable to set up and maintain capability for all types of analyses (this saves customers the inconvenience of sending multiple samples and ensures the laboratory can still secure a portion of the work, and hence income).

In order to estimate the investment required to perform different types of analyses, a calculation of the cost to perform the proximate analyses as well as the cost to implement other types of analyses should be made. The cost calculation should include costs for facilities, equipment, personnel and consumables.

2.2.4 Market analysis for available laboratory services

The next step in the process is to examine the current market situation for performing analytical services, and the opportunities for changing this situation. The success of the new laboratory depends mostly on the opportunity to take over part of the existing market, so it is therefore important to investigate how much flexibility currently exists. This assessment should focus on the following issues:

- Estimation of the number and type of laboratories already present in a specific area. For the characterization of the laboratories, the division as earlier mentioned (Section 2.2.1) can be used. This analysis will lead to identification of laboratories that can be seen as competitors for the potential new laboratory.
- Identification and assessment of the relationship between the laboratories and the clients. This should give information on the amount different clients will spend on obtaining laboratory services. The relationship between clients and a laboratory will be based on the quality and promptness of services provided to a client. This will also determine their loyalty and personal preference towards a company. Company policy may dictate the flexibility each potential client will have to make a change in their out-sourcing of laboratory business. Some of this information may be commercially sensitive and difficult to obtain.
- Visiting potential clients as part of a Public Relations (PR) exercise can be valuable in establishing contacts in the industry and establishing an indication of the amount of analysis work that could be available and the type of service expected by the clients (e.g. are they dissatisfied with their current suppliers, and if so, for what reason?).

2.2.5 Evaluation and decision-making

The last step in the creation of the business plan is to bring together all the information collected in the previous steps to make an evaluation of the different options. For calculating profitability, the evaluation should estimate the potential revenues and costs. This should be done for the different types of laboratories in Section 2.2.1 above.

The calculation of the potential revenue should be based on the results of the market analysis for each of the clients (Section 2.2.2) and their relationship with the existing laboratories (Section 2.2.4). The first step in this calculation is to gather information on the different types of analysis and the amount a client pays to each of the laboratories in the region. The second step is to estimate how many potential clients would be willing to make a change and to procure services from the new laboratory, and estimate this potential revenue. This likely change can be expressed in terms of a probability factor with a scale from 0 to 1, with 0 being low probability of change to the new laboratory and 1 being a high probability of change. The total revenue is estimated by multiplying the amount spent by the probability factor. As there are a number of factors beyond the control of the laboratory, this figure will only provide an estimate, as an exact figure is difficult to derive. In general, clients that use more than one laboratory are more willing to switch to a new laboratory, compared with those that use only one, provided the new laboratory meets all their analytical needs.

Focusing on a new market can be much more profitable than looking at just the existing market, but it means more uncertainty because it depends on future plans of potential clients. Therefore, it is preferable to focus in the first instance on potential customers within the present market.

The total costs should also be calculated for all the different types of analyses. A good approach is to calculate firstly the costs to perform only the proximate analyses, followed by an estimation of the additional costs to perform other, more advanced, analyses. Costs will vary widely between countries, depending on freight costs, currency exchange rates, availability of suitable equipment or of high grade consumables, and labour costs.

The expected profits for the different options can be calculated from the predicted total revenue and costs. Calculation of profit, however, is based on various input variables that contain uncertainties. A sensitivity analysis can easily be performed by changing the value of input variables and assessing the effect on the calculated profit. This analysis should be limited to those variables that contain the highest uncertainty and therefore have the greatest effect on the accuracy of the calculation. Some examples of such variables are the probable costs and the prices of analyses charged by other laboratories. The evaluation of this uncertainty can be performed by estimating the difference between the predicted profit and a non-profit situation.

The new laboratory should be careful if using the market price for calculating the revenues because it will be competing with other, established laboratories. The new laboratory should avoid using prices that are too low in order to capture a part of the market: a reduction in prices might increase the volume of work, but the consequences could be a decrease in profit as a result of prices with too little profit margin. Also, it could lead to a general 'price war' amongst laboratories, with destructive consequences. In addition, laboratories might find it difficult to raise prices after a certain period.

The final result of these calculations are values for the profitability of the laboratory under different conditions, expressed as an average value with a confidence interval. The range of this confidence interval reflects mainly the uncertainty in the estimation of the potential revenues. The profitability is often expressed as Return on Investment (ROI) which is related to the investments required. The decision-makers and investors will use these values to make decisions regarding the investment in the new laboratory. Depending on the type of laboratory (see Section 2.2.1), however, the decisions could be made in different ways:

- **Stand-alone commercial laboratories.** The decision will be based purely on the level of profitability and its confidence interval. The uncertainty expressed in the range of the confidence interval will positively affect the margin of profit that investors demand.
- **Laboratories integrated with feed producing units.** The decision should primarily be influenced by the profit a feed producer is likely to make from the feed manufacturing activity. In this case, an alternative approach is to compare the costs of the laboratory against those for outsourcing the analyses to be undertaken for effective running of the feed manufacturing unit. A laboratory integrated into the feed manufacturing unit has several benefits, such as a quick turnaround time, not dependant on any outside laboratory, and better quality control of the products. These benefits should also be quantified and taken into consideration when making a decision.
- **Laboratories integrated with research institutes.** The decision will primarily be influenced by the additional value the laboratory brings to the research conducted in the institute. Although such laboratories can also perform analyses for commercial clients, practice shows that these laboratories often have difficulty competing with stand-alone commercial laboratories, mainly due to the high throughput of the latter and thus lower unit costs per analysis. Nevertheless, research laboratories should also try to generate additional revenue by attracting commercial clients. The additional

revenue could help finance new instruments and analyses required to meet fast moving research needs.
- **Laboratories integrated with educational organizations.** The decision should primarily be based on enhancing the quality of education and producing graduates and researchers having the required skills for addressing the challenges of the industry and capable of contributing substantially to cutting-edge science. As for the laboratories integrated with research institutions, the educational laboratories should try to attract commercial clients; however, they should be cautious and not assign such work to staff lacking the appropriate training and competence, such as trainees and students.
- **Government or reference laboratories.** The decision to create this type of laboratory should, in contrast to the other laboratories, be based not on commercial benefits but purely on regulatory reasons. Its central position and high quality demands require a large investment in personnel, equipment and facilities, that should be funded by public money to guarantee the independence of the laboratory and avoid conflict with commercial interests.

Chapter 3
Setting up and running the laboratory

3.1 INTRODUCTION
After a positive decision is made to set up a laboratory, actions need to be taken to put into practice the business plan developed in Chapter 2. The setting up of a new laboratory involves:
- Selection or construction of a building and facilities required for various analyses (Section 3.2.3).
- The analytical process and an organizational structure to facilitate this (Sections 3.2.1 and 3.2.5).
- Selection of analyses to be performed (Section 3.2.2).
- Selection and purchase of equipment (Section 3.2.4).
- Attracting and maintaining qualified staff (Section 3.4).
- Establishing standard operational procedures (SOPs), i.e. formally written controlled documents outlining all the steps for each of the methods the laboratory decides to undertake (Section 3.3).

In most cases a Project Steering Group (PSG) is constituted to bring the above six stages to a satisfactory completion within a fixed time schedule. This group should contain a technical expert with experience in a feed analysis laboratory, a laboratory manager with experience in quality assurance (or Quality Assurance (QA) manager), and personnel from finance, procurement and management areas. Throughout the process of setting up the laboratory, good communication among the members of the PSG is vital to ensure success.

This chapter will focus on the realization of an analytical chemical laboratory. For microbial determinations, which are also important analyses in animal feedstuffs, special facilities and equipment are needed, such as clean and separate rooms to avoid contamination, and flow cabinets to work under safe conditions. A detailed description of these requirements is outside the scope of this manual. Special facilities and equipment to conduct laboratory research in the field of animal nutrition, such as *in vitro* or *in situ* investigations, are also outside the scope of this manual.

3.2 PHYSICAL REALIZATION OF THE LABORATORY
The physical realization of an operational laboratory involves the construction of the laboratory, purchase of equipment and appointment of personnel. All these issues are related to the analytical work, and more specifically to the methods, that the laboratory intends to conduct. The choice of methods is therefore a critical step (see Figure 3.1). This section starts with an overview of the analytical process and available methods, followed by their implications for construction or selection of buildings and facilities, purchase of equipment and putting in place an organizational structure with defined responsibilities for the personnel involved.

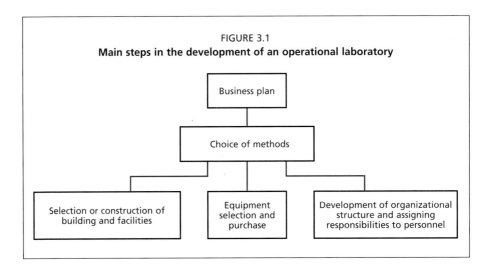

Technical knowledge and experience should not dictate 'Choice of methods'. It should be directed only by commercial or scientific criteria, as stated in the business plan. If some technical knowledge and skills are not available, then appropriate technical staff can be recruited.

3.2.1 The analytical process
The analytical process is the foundation of the laboratory, and ensures that procedures and protocols are followed to consistently produce results of a high standard that meet international QA requirements. The various stages of the analytical process are described in Figure 3.2.

This process starts with the receipt of samples and a request from the client for the analyses to be performed. On receipt of the samples and appropriate sample preparation

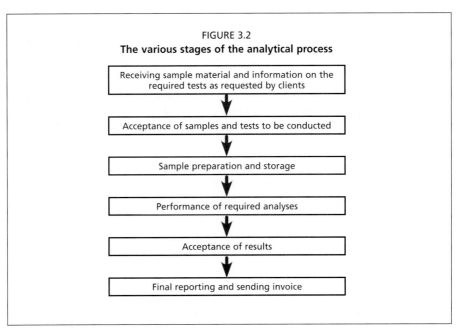

and storage, analyses can begin. The results of these tests are collated and checked, and once approved by an authorized person, a final report, including an invoice, is sent to the customer. It is important to make sure that all requests from clients have been noted as well as the most suitable method chosen (if alternatives are available such as fat with or without prior hydrolysis). Responsibility for checking these details should be clearly defined.

Sample materials are stored in the laboratory for a fixed time, e.g. one month, from completion of analyses and either returned (on request), discarded or destroyed.

3.2.2 Methods to be made operational

The methods used in a feed analysis laboratory can be divided into proximate and specific instrumental analyses. The proximate analyses are used for characterization of feeds based on macro-nutrients, such as dry matter, protein, fat, fibre and ash, whereas the advanced (instrumental) analyses focus on specific components, like individual minerals, amino acids and fatty acids. A detailed description of these methods can be found in a previous FAO publication (FAO, 2011). Since the level of analytical skills required differs for each method, it is vital that each technician is 'signed off' as being trained and capable of performing each specific method, and this should be recorded in the training file.

Near-Infrared Spectrometry (NIR) is sometimes mentioned as an alternative method to estimate the nutritional composition of feeds, and needs only a power supply and a grinding machine, and possibly also an internet connection. However, laboratories should regard NIR only as an additional method that should be based on the results they have determined by traditional methods, as described in this chapter.

The main analyses and the steps involved in conducting these are given below (for more specific method details, see FAO, 2011).

For personnel, technical skill levels should be categorized at three levels:

- **Basic analytical skills.** Familiarity with the use of balances, sample mixing and sub-sampling. No specific background or education required other than an attentive attitude and attention to detail.
- **Medium analytical skills.** Understanding of basic chemical reactions and the principle of the method being applied; awareness of laboratory safety when working with solvents and strong acids and bases; computer competency; use of gas cylinders; spectrophotometer; use of analysis-specific equipment such as Fibertec or Soxtec; and bomb calorimeter. Laboratory experience is essential.
- **High analytical skills.** Specific instrument training, i.e. High Performance Liquid Chromatography (HPLC); Ultra HPLC; Gas Chromatography (GC); Gas Chromatography-Mass Spectrometry (GC-MS); Liquid Chromatography-Mass Spectrometry (LC-MS); and Inductively coupled plasma atomic emission spectroscopy (ICP-AES); associated software programs; able to maintain required instrumentation. Ability to make independent decisions regarding peak identification and its area. Awareness of laboratory safety requirements when working with toxic and carcinogenic compounds. Laboratory experience is essential, along with relevant University or recognized technical qualifications.

Typical tests in a feed analysis laboratory require specific facilities and skills. These are listed in Table 3.1.

TABLE 3.1
Typical tests in a feed analysis laboratory and their technical requirements

Parameter	Description
Sample preparation	
Description	Sample preparation is essential for sub-sampling of material prior to a determination.
Activities	Drying and grinding.
Equipment	Low temperature oven dryer (60–70 °C) or freeze dryer; splitter; mill; sieves.
Facilities	Two- or three-phase electric power; exhaust system.
Personnel	Basic analytical skills.
Dry matter analysis	
Description	Dry matter is by definition the part of the sample that remains after drying at 103 °C.
Activities	Weighing and drying.
Equipment	Analytical balance (0.1 mg); forced-air drying oven (at least 110 °C); desiccator.
Facilities	Granite (or similar) table for balance stability; an oven connected to an exhaust system.
Personnel	Basic analytical skills.
Crude ash	
Description	Crude ash is by definition the part of the sample that remains after incineration at 550 °C.
Activities	Weighing and incineration.
Equipment	Analytical balance (0.1 mg); desiccator; muffle furnace.
Facilities	Connection to exhaust ventilation system for muffle furnace; granite (or similar) table for balance stability.
Personnel	Basic analytical skills.
Ash insoluble in acid (sand)	
Description	Ash insoluble in acid is the ash that remains after boiling in strong acid.
Activities	Weighing, boiling and incineration.
Equipment	Analytical balance (0.1 mg); desiccator; muffle furnace; heating and reflux equipment.
Facilities	Granite (or similar) table for balance stability; a fume hood connected to an exhaust system.
Personnel	Basic analytical skills.
Crude protein	
Description	The term 'crude protein' refers to measuring the total nitrogen content and to calculate the protein content by multiplying the nitrogen content by an appropriate conversion factor (usually ×6.25). If an alternative method such as the summation of amino acids is used, the term 'crude protein' should not be used. Two methods, Kjeldahl and Dumas, are available for nitrogen determination.
Crude protein – Kjeldahl method	
Description	Nitrogen is converted into ammonia which is absorbed in boric acid and titrated against a standard acid.
Activities	Weighing, digestion, distillation and titration.
Equipment	Analytical balance (0.1 mg); digestion unit; distillation unit; titration unit.
Facilities	Granite (or similar) table for balance stability; fume hood connected to an exhaust system.
Personnel	Medium-level analytical skills.

(cont.)

TABLE 3.1 *(cont.)*

Parameter	Description
Crude protein – Dumas method	
Description	With complete combustion of sample at 950 °C in the presence of oxygen, nitrogen is converted to a gaseous state and reduced to N_2, followed by measurement in a thermal conductivity cell.
Activities	Weighing, combustion to N_2, and measurement.
Equipment	Analytical balance (0.1 mg); Dumas apparatus.
Facilities	Granite (or similar) table for balance stability; helium and oxygen gas supply (high purity; 5.0).
Personnel	Medium- to high-level analytical skills.
Crude fat	
Description	Crude fat is by definition the non-polar extractable fraction of the sample. The extraction can be performed with or without prior acid hydrolysis, both being complementary methods. The laboratory should offer both options.
Activities	Weighing, hydrolysis, filtration, extraction and drying.
Equipment	Analytical balance (0.1 mg); units for heating, filtration, extraction and refluxing; forced-air drying oven or vacuum oven (preferable).
Facilities	Granite (or similar) table for balance stability; fume hood connected to an exhaust system.
Personnel	Medium-level analytical skills.
Fibre analysis	
Description	Fibre analysis is based on boiling of the sample in a special detergent solution and measurement of the remaining organic fraction. There are two methods available, both are based on digestion of feeds in detergent solution: (1) digestion of feed directly in the detergent solution and filtration *using crucibles* (this is the official standard method); and (2) digestion of sample whilst in a *nylon bag* and then washing the bag containing the digested sample to make it detergent free.
Fibre analysis – Crucible-based filtration method	
Description	Digestion of feed directly in the detergent solution and filtration *using crucibles* (this is the official standard method).
Activities	Weighing, boiling, filtration, drying and incineration.
Equipment	Analytical balance (0.1 mg); hot plate; reflux and filtration unit; forced-air drying oven; muffle furnace; crucibles.
Facilities	Granite (or similar) table for balance stability; fume hood connected to an exhaust system.
Personnel	Medium-level analytical skills.
Fibre analysis – Nylon bag-based method	
Description	Digestion of sample whilst in a *nylon bag* and then washing the bag containing the digested sample to make it detergent free.
Activities	Weighing, boiling, washing, drying and incineration.
Equipment	Analytical balance (0.1 mg); ANKOM apparatus; forced-air drying oven; muffle furnace.
Facilities	Granite (or similar) table for balance stability; fume hood connected to an exhaust system.
Personnel	Medium-level analytical skills.
Starch	
Description	Starch can be measured by the classical Ewers method or with an enzymatic method. The enzymatic method can be used for all sample types and is therefore preferable.
Activities	Weighing, extraction, incubation, dilution, and spectrometric measurement.
Equipment	Analytical balance (0.1 mg); temperature-controlled water bath; autoclave (optional); suction unit; volumetric equipment; spectrometer.
Facilities	Granite (or similar) table for balance stability; fume hood with vacuum system.
Personnel	Medium-level analytical skills.

(cont.)

TABLE 3.1 *(cont.)*

Parameter	Description
Reducing sugars	
Description	Reducing sugars contain the most important sugars, including glucose, fructose and sucrose. Determination is based on the Luff-Schoorl principle.
Activities	Weighing, incubation, dilution, and titration or spectrometric measurement.
Equipment	Analytical balance (0.1 mg); temperature-controlled water bath; volumetric equipment; titration unit or a spectrophotometer (depending on method).
Facilities	Granite (or similar) table for balance stability.
Personnel	Medium-level analytical skills.
Gross energy	
Description	Gross energy represents the total energy value of the sample and is measured by bomb calorimeter.
Activities	Weighing, instrumental measurement, and titration.
Equipment	Analytical balance (0.1 mg); bomb filling system; bomb calorimeter; titration unit.
Facilities	Granite (or similar) table for balance stability; oxygen supply; fume hood connected to an exhaust system.
Personnel	Medium-level analytical skills, with some experience.
Minerals	
Description	Minerals are generally measured by spectrometric methods following incineration and hydrolysis.
Activities	Weighing, incineration (optional), acid digestion, dilution, spectrometric measurement, and Atomic Absorption Spectroscopy (AAS) or Inductively Coupled Plasma-Atomic Emission Spectroscopy (ICP-AES) instrumental measurement.
Equipment	Analytical balance (0.1 mg); muffle furnace (550 °C) with connection to an exhaust system (optional); heating plate or digestion unit (250 °C); volumetric equipment; AAS and spectrometer or ICP-AES.
Facilities	Granite (or similar) table for balance stability; fume hood connected to vacuum system; acetylene and air supply for AAS, or argon supply and three-phase current for ICP-AES; high purity water.
Personnel	Medium- to high-level analytical skills.
Amino acids (excluding tryptophan)	
Description	The standard method for the determination of amino acids is based on the hydrolysis of protein to amino acids using a strong acid with or without previous oxidation, followed by chromatographic separation and detection after derivatization.
Activities	Weighing, oxidation (optional), hydrolysis, evaporation, and chromatographic measurement.
Equipment	Analytical balance (0.1 mg); hydrolysis unit; oven (110 °C); evaporation equipment; HPLC or dedicated amino acid analyser.
Facilities	Granite (or similar) table for balance stability; fume hood connected to an exhaust system; cold water supply; compressed air for autosampler (optional); helium supply for degassing buffer solutions; high purity water.
Personnel	High-level analytical skills.
Amino acids – tryptophan	
Description	Determination of tryptophan is based on an alkaline hydrolysis followed by chromatographic separation.
Activities	Weighing, hydrolysis, and chromatographic measurement.
Equipment	Analytical balance (0.1 mg); air-forced oven dryer; HPLC system attached to a UV- or fluorescence detector.

(cont.)

Setting up and running the laboratory

TABLE 3.1 *(cont.)*

Parameter	Description
Facilities	Granite (or similar) table for balance stability; fume hood connected to an exhaust system; compressed air for autosampler (optional); helium supply for degassing buffer solutions; high purity water.
Personnel	High-level analytical skills.
Fatty acids	
Description	The standard method for fatty acids is based on isolation and derivatization, followed by gas chromatographic separation.
Activities	Weighing, derivatization, and chromatographic measurement.
Equipment	Analytical balance (0.1 mg); air-forced oven dryer; GC system attached to a Flame Ionization Detector (FID).
Facilities	Granite (or similar) table for balance stability; fume hood connected to an exhaust system; cool water supply; compressed air for autosampler (optional); helium supply as carrier gas; hydrogen and compressed air for the FID detector.
Personnel	Medium- to high-level analytical skills.
Vitamins	
Description	Determination of individual vitamins is based on extraction, followed by clean up, concentration if needed, and chromatographic measurement.
Activities	Weighing, extraction, purification, and chromatographic measurement.
Equipment	Analytical balance (0.1 mg); temperature-controlled water bath; unit for solid phase extraction; volumetric equipment; HPLC system including UV- or fluorescence detector.
Facilities	Granite (or similar) table for balance stability; fume hood connected to an exhaust system; compressed air for autosampler (optional); helium for degassing elution solution; high purity water.
Personnel	High-level analytical skills.
Mycotoxins	
Description	Mycotoxins are undesirable substances produced by fungi (moulds). These present a potential danger to animal and human health. The maximum levels are nationally and internationally regulated. The different methods are based on extraction, purification, chromatographic separation and detection.
Activities	Weighing, extraction, purification, and chromatographic measurement.
Equipment	Analytical balance (0.1 mg); temperature-controlled water bath; unit for solid phase extraction; volumetric equipment; HPLC system including the possibility for pre- or post-column derivatization and fluorescence detection.
Facilities	Granite (or similar) table for balance stability; biological safety cabinet; fume hood connected to an exhaust system; compressed air for autosampler (optional); helium for degassing elution solution; high purity water.
Personnel	High-level analytical skills
Pesticides	
Description	Pesticides are undesirable substances whose maximum levels are defined in national and international regulations. These regulations demand a low detection limit and positive identification of the pesticides, which is achieved by using mass spectrometric detection. The methods are based on extraction, purification, derivatization, chromatographic separation and identification.
Activities	Weighing, extraction, purification, derivatization, and chromatographic measurement.
Equipment	Analytical balance (0.1 mg); temperature-controlled water bath; unit for solid phase extraction; volumetric flasks; GC-MS, including a databank for identification of the individual components.
Facilities	Granite (or similar) table for balance stability; fume hood connected to an exhaust system; compressed air for autosampler (optional); helium as a carrier gas.
Personnel	High-level analytical skills.

3.2.3 Building and facilities

The next step in the process is the choice of location, and construction of the laboratory, including facilities. If connected to a large organization, such as a feed manufacturer, the location of the laboratory is generally pre-determined. If there is a choice as to where the laboratory could be located, the presence of some basic requirements, including good infrastructure (i.e. road system) and assured and uninterrupted power and water supply, are crucial elements in the selection of the location. The location should also be chosen with a view to possible expansion plans in the future, and ensuring that local legislation allows the construction of a laboratory on that site.

After confirming the location, a decision regarding the type of building is required. In some cases there will be the possibility of renovating an existing building, which might appear to be a financially attractive option. In practice, however, refurbishment of an old building is often very expensive and complex, which may lead to a sub-optimal situation requiring possible future expansion of the laboratory. Construction of a new laboratory building is therefore the preferred option. For either building option, advice from an architectural company specializing in laboratory design will be invaluable, especially with respect to conforming to the relevant local and national health and safety requirements.

A good approach while designing the laboratory is to base it on the methods to be used, and possible future analytical requirements. Allow for extra power points, water outlets and fume hoods wherever possible, and computer network access points as appropriate.

Suggested divisions or sections for a feed analysis laboratory are as follows:

- **Sample registration.** An area convenient for clients and couriers to deliver samples. Here the samples are logged into a booking system, given a unique identification number, and the required analyses selected. Samples are then passed on to the sample preparation section.
- **Sample preparation.** An area where sub-sampling, blending, grinding and, if necessary, pre-drying occur. Pre-drying of samples is necessary if the laboratory deals with products with a moisture content >15%, such as silage samples or manure. For pre-drying, freeze drying is preferable to oven drying at 70 °C as there is less damage to the structure of the sample and to the heat-sensitive components such as amino acids and vitamins. Weighing, drying and incineration may also be conducted in this area. Keep heat-generating equipment such as ovens and furnaces in one area, and if necessary an air extraction unit can be utilized to remove odour as well as excess heat.
- **Digestion, filtration, distillation, dilution and titration.** Keep a separate area for acid use and storage. These areas require access to water and should be close to glassware supplies, balances, fume hoods and chemical supplies.
- **Extraction, derivatization and dilution.** Keep a separate area for solvent use and storage. This area requires access to water and should be close to glassware supplies, balances, fume hood and chemical supplies.
- **Instrumental measurements.** Depending on the choice of instruments, specialized conditions may be required, such as air conditioning, dust-free work area, no direct sunlight, special type of power supply, gas supply, etc.

The separation of laboratory space to perform the above activities is primarily required to avoid cross-contamination with undesirable substances and to maximize the use of

Setting up and running the laboratory

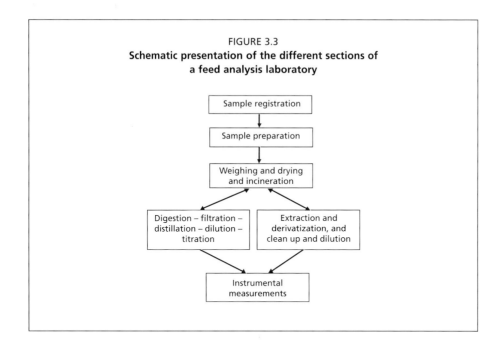

FIGURE 3.3
Schematic presentation of the different sections of a feed analysis laboratory

space and staff time. Sample preparation produces dust and noise, and should be physically separated from other activities. Weighing, drying and incineration are mostly linked to each other and do not involve working with chemicals. Traditional wet chemistry is split into two sections based on the type of chemicals (i.e. strong acids and bases versus organic solvents), and these two sections should be physically separated to avoid health and safety issues with mixing of flammable solvents and corrosive chemicals. Also, keep separate fume hoods for each of these sections. Sensitive instruments are placed in a clean environment, away from other activities. If mycotoxins are to be analysed there will be a requirement for a biological safety cabinet to be available in the laboratory. Figure 3.3 shows a schematic presentation of different sections of a feed analysis laboratory.

For maximum efficiency, after the samples have been registered and the analyses assigned to technical staff, laboratory activities should be separated into five different sections, each with different requirements.

Section 1. Sample preparation
Area   ca 24 m^2
Equipment and related items   Grinding machine and sieves
                              Brushes for cleaning sieves and grinder
                              Three-phase electric power
                              Cubicles connected to a ventilation system for grinding
                              Drying ovens
                              Freeze drier
                              Compressed air
                              Network connection or data transfer for automatic weight recording from balances into a spreadsheet (optional).

Furniture and related items	Work table/bench
	Storage facilities at low temperature (for example refrigerator, freezer) as well as at room temperature
	Quarantine facility for imported or potentially infectious samples.
Safety items	Dust masks
	Safety glasses and ear protection
	Hand washing facilities
	First aid kit

Section 2. Weighing room (including drying and incineration)

Area	ca 36 m^2
Equipment and related items	Granite (or similar) weighing tables
	Balances
	Cubicles connected to an exhaust system (care is needed to ensure there is no draught produced which could affect the balance accuracy)
	Three-phase power (for muffle ovens)
	Weighing balances
	Network connections or data transfer for automatic weight recording from balances into spreadsheet (optional).
Furniture and other items	Work tables and benches, including chairs
	Storage facilities
	Glassware and other standard laboratory items such as desiccators, tongs, spatulas, beakers and crucibles.
Safety items	Laboratory coat
	Dust mask and gloves
	Heat resistant gloves
	Fire extinguisher
	First aid kit

Section 3. Digestion room

Area	ca 48 m^2
Equipment and other items	Digestion heating blocks
	Fume hoods connected to an exhaust system (suitable for acid fumes)
	Water and gas supply
	Vacuum facilities
Furniture and related items	Work tables and benches, including chairs
	Safety cabinets for storage of chemical solutions (acids and bases to be stored separately) and chemicals
	Glassware, including beakers, crucibles, dispensers, pipettes and measuring cylinders

Safety	Laboratory coat
	Gloves
	Safety glasses
	Eye wash station
	Fire extinguisher
	Containers for chemical waste
	Chemical spill kits
	First aid kit

Section 4. Extraction room

Area	ca 48 m^2
Equipment and related items	Fume hoods connected to an exhaust system for solvent extraction
	Filtration unit
	Reflux system
	Acid concentrator
	Rotary evaporator
	Fibre and fat extraction systems
	Centrifuges
	Water and gas supply.
Furniture and other items	Work tables and benches, including chairs
	Safety cabinets for storing chemical solutions and organic solvents and chemicals
	Other laboratory items such as glassware, dispensers, transfer pipettes, crucibles, extraction thimbles, etc.
Safety items	Laboratory coat
	Gloves
	Safety glasses
	Eye wash station
	Fire extinguisher
	Containers for chemical waste
	First aid kit

Section 5. Instrument room

Area	ca 48 m^2
Equipment and related items	Gas Chromatograph with or without Mass Spectrometer detector (GC, GC–MS)
	High Performance Liquid Chromatograph (HPLC) with or without Mass Spectrometer detector (LC-MS) and Ultra HPLC
	Atomic Absorption Spectrometer (AAS)
	Inductively Coupled Plasma Atomic Emission or Mass Spectrometer (ICP-AES or ICP-MS)
	Amino acid analyser

	Spectrophotometer
	Vortex
	Ultrasonic water bath
	Filtering system
	N-analyser
	Uninterrupted power supply
	Water and gas supply
	Air conditioning
	Purified water system for chromatography work
	Network connections to allow direct laboratory access to data generated from the equipment
Furniture and related items	Work tables and benches, including chairs
	Equipment manuals
Safety items	Laboratory coat
	Gloves
	Safety glasses
	Eye wash station
	Fire extinguisher
	First aid kit
Note:	If mycotoxins or residues are to be analysed, all processes that pose a risk to the health of operators should be performed in a biological safety cabinet.

Additionally, the laboratory should have the following separate facilities:

Section 6. Sample storage room

Area	ca 24 m^2
Equipment and related items	Air conditioned or cool dark room
	Freezer
	Refrigerator
Furniture and related items	Storage possibilities such as shelves and cupboards for sample material
Safety	Fire extinguisher

Section 7. Dish washing and drying facility

Area	ca 12 m^2
Equipment and related items	Hot-air oven (110 °C)
	Dishwasher and autoclave (both are optional)
Furniture and related items	Water supply and drain
	Tiled floor and walls
	Work table and bench
	Storage facilities

Section 8. Administration or office area

Area	ca 20 m²
Facilities	Computers
	Air conditioning unit
	Scanner
	Photocopier
	Fax
Furniture and related items	Work tables, including chairs
	Client information
	Storage facilities for results

Section 9. Welfare and rest area
A suitable clean area should be available for staff to take breaks, eat meals etc. This should include toilets and showers; and a changing area should be available for staff to change and store their clothing. Appropriate laundry facilities may be required for laboratory coats.

A total of ca 236 m² is required to conduct the analytical work in a safe and efficient manner. The exact area of the laboratory depends on the number of instruments and staff employed. The space mentioned above will be sufficient to accommodate all basic equipment for the analyses and eight technical staff, this space can easily deal with a few thousand samples each year. If the laboratory has fewer technical staff and is dealing with a small number of samples the area required can be reduced accordingly, to a minimum of 40 m² to conduct the basic proximate analyses (e.g. dry matter, ash, crude protein, crude fibre and crude fat) and the spectrometric determination of macro-elements. In some laboratories, the Instrument room is partitioned based on the type of equipment, such as chromatographic (e.g. HPLC, GC) and spectrometric equipment (e.g. AAS and ICP); a partition to isolate the 'noisy equipment' such as centrifuges and vacuum pumps is strongly advised.

It is preferable that the various sections of the laboratory are located on the same floor. This facilitates the safe transport of samples and chemicals. It is also more efficient for technical staff not to have to carry samples and potentially dangerous chemicals to different areas along corridors, etc. The different sections should be arranged in a logical manner to minimize distances and to avoid conflict in activities. A suggested option is to cluster the sections as follows:
- Sections 2, 3, 4 and 5 close to each other, as they are all part of the analytical process.
- Sections 1 and 6 next to each other so that the samples can be conveniently stored after appropriate preparation and analysis.
- Sections 7 and 8 should be placed separately to guarantee clean working conditions for both the activities.
- Section 9 should be separate from 1 to 8, perhaps located adjacent to office and other 'clean' facilities.

All this information will help the PSG to draw up a draft design for the future laboratory. To finalize this design, the following steps are needed.
- Ensure the draft design, including the technical specifications, such as ventilation, is evaluated by an external consultant, such as an architectural company specializing in laboratory design. The technical specifications will vary from country to country,

depending on legislative requirements. Local construction companies as well as the local regulatory agency may be able to recommend suitable companies to contact for this purpose. It is advisable to use their expertise to improve the design, and it may be prudent to involve them at an early stage of the process.
- Evaluate the draft design, focusing on safety and environmental dimensions. As stated previously, the laboratory should guarantee safe and healthy working conditions for all staff and avoid pollution or contamination of the surrounding environment. Important issues are storage of chemicals; safety while working with explosive or toxic gases and chemicals; ventilation requirements; and waste control and disposal. In most countries, there is legislation on these issues. The laboratory should ensure that the required facilities are provided to enable work to take place as per regulatory requirements. Consultation with national authorities on this issue can be very helpful.
- Contact different construction companies to get estimates for the cost of constructing the design. These companies should have experience and expertise in the construction of laboratories. They must also be familiar with health and safety regulations, and special materials required (for example acid- and alkaline-resistant bench tops and floors).

The final design, including the projected cost for the completed project, will need to be approved by the decision-makers, prior to starting the tendering and construction stages. This process should be carefully monitored by members of the PSG. At each stage of the design process, check that the basic constituents remain as required and meet the required quality and safety criteria.

3.2.4 Equipment

The equipment for the laboratory depends on the analyses it intends to conduct, as listed in Sections 3.2.1 and 3.2.2.

For proximate analyses, laboratories have the option of performing the analyses manually or to use specialized equipment. The advantages of the manual methods are lower costs for equipment and generally lower maintenance costs. These methods are, however, more laborious and time consuming. A list of equipment required is given in Table 3.1. Costs given in Tables 3.2 to 3.4 are indicative, based on typical prices in the Netherlands in 2013.

If an equipment item is utilized for several different methods, such as an analytical balance, more than one may be required to ensure maximum efficiency of staff time.

The list in Table 3.2 shows that at least € 35 000 is needed in order to set up a basic laboratory able to perform the proximate analyses by manual methods. An alternative to using manual methods is to purchase specialized equipment for the different determinations, as described in Table 3.3. This approach requires a much higher financial investment, and is only justifiable if a high sample throughput can be guaranteed.

For specific instrumental analyses, there are no alternatives other than to use dedicated equipment. The cost for this equipment varies between suppliers and the technical specifications. Table 3.4 gives indicative prices for the most commonly used equipment for each type of analysis.

The laboratory also requires standard consumables and tools to facilitate the analytical process.

TABLE 3.2
Equipment needed to manually perform proximate analyses

Equipment	Analyses	Costs (Euro)
Analytical balance	All	1 000
Drying oven	Dry matter, fibre and fat	2 000
Muffle furnace	Ash, acid insoluble ash, and fibre	2 500
Heating device	Acid insoluble ash, fibre, and fat	1 500
Extraction and reflux unit	Fat	3 000
Digestion unit	Nitrogen	10 000
Distillation unit	Nitrogen	3 500
Automatic titration unit	Nitrogen and sugar	350
Filtration unit	Fibre and fat	2 000
Temperature controlled water bath	Sugar and starch	2 000
Spectrometer	Sugar and starch	3 000
Volumetric equipment	Sugar and starch	1 500

Notes: The costs do not include value added tax, and are approximate in 2013.

TABLE 3.3
Specialized equipment needed for proximate analyses

Equipment	Analyses	Unit comprises	Costs (Euro)
Kjeldahl determination	Nitrogen	Distillation and titration unit	35 000
Fat determination	Fats and oils	Extraction apparatus, semi-automatic	20 000
Fibre determination	Neutral and acid detergent fibre, lignin	Boiling and filtration apparatus, semi-automatic	25 000
Dumas determination	Nitrogen (total combustion method, alternative to Kjeldahl)	Auto-sampler, balance, Dumas apparatus	45 000
ANKOM fibre determination	Neutral and acid detergent fibre (nylon bag method, alternative fibre determination)	Analytical balance (0.1 mg), ANKOM apparatus, forced-air drying oven, and muffle furnace	15 000
Bomb calorimeter	Gross energy	Balance, bomb calorimeter	50 000

Notes: The costs do not include value added tax, and are approximate in 2013.

TABLE 3.4
Dedicated equipment required for instrumental analyses

Equipment	Analysis	Costs (Euro)
Atomic Absorption Spectroscope (AAS)	Minerals	35 000
Inductively Coupled Plasma Analysis (ICP)	Minerals including P	50 000
Gas Chromatograph-Flame Ionization Detector (GC-FID)	Fatty acids	30 000
Gas Chromatograph-Mass Spectrometer (GC-MS)	Pesticides	50 000
High Performance Liquid Chromatograph (HPLC)	Amino acids, vitamins	30 000

Notes: The costs do not include value added tax and are approximate in 2013.

In some cases there may be alternative solutions. Thus for GC work the gas supply has traditionally been provided by gas cylinders purchased from a bulk gas supplier, but it may be more economical to install a hydrogen generator, which then requires water and power. Selection of equipment needs to be based on set criteria, including technical specifications, such as detector, wavelength range, centrifuge rpm (*g* value), price, etc., and also the level of post-purchase service available. Service contracts are available and should be investigated thoroughly prior to purchase. Always ensure that a full service manual is provided (including electrical diagrams). For optimum performance, regular maintenance is critical, provided either by company staff or external entities.

3.2.5 Organizational structure and responsibilities of personnel

For the laboratory, the number of personnel and their educational and experience levels depends on the analyses to be offered, the methods chosen and the expected sample throughput. The first step is to create an organized structure for the laboratory and to define the activities to take place in the laboratory, as illustrated in Figure 3.4.

Technical staff may also be referred to as 'Technicians', 'Analysts' or 'Scientists'.

The horizontal division of the laboratory into different sections depends on the type of analyses it intends to perform. Similar types of analyses, such as proximate or chromatography (GC, HPLC) are grouped into one section. Ideally, a senior technician will be responsible for each area, such as proximate analyses, with 2 or 3 technicians to rotate around the various analyses within their 'section'. This enables an overall knowledge to be gained of the working requirements for each area of the laboratory. It is important to keep the job interesting and challenging for staff and to avoid monotonous, repetitive work where ever possible. The vertical division of the laboratory reflects the different positions and responsibilities within the organization. As a guide, some typical positions can be identified:

- The Laboratory Manager is responsible for the whole laboratory and the development of its strategic plan. A key part of this position is the external communication with clients and potential clients, as well as full responsibility for results reported to clients. Management systems must be put in place to ensure reliable data are produced and that the reporting of this data is thoroughly checked prior to releasing reports.
- The Quality Assurance (QA) Manager is responsible for quality assurance within the laboratory, and should have an independent position. The QA Manager may also have responsibility for Health and Safety Management, or Environmental Management

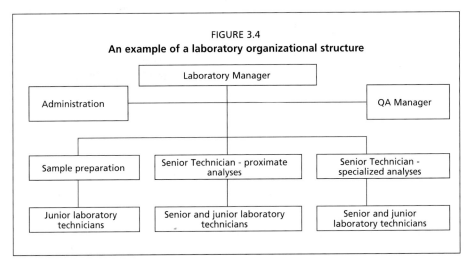

FIGURE 3.4
An example of a laboratory organizational structure

Setting up and running the laboratory 27

within the laboratory, or both, or they may be the responsibility of a separate Health and Safety Manager and an Environmental Manager.
- The Section Head or Senior Technician (Proximate analyses) is responsible for the daily organization of the analytical process, ensuring that daily and weekly deadlines within their section are met; quality control for each batch of testing meets requirements and is recorded; staff training is up-to-date; and that there are sufficient staff to meet the workload requirements. Maintaining stocks of the necessary chemicals and consumables are also the responsibility of the senior technician, who should inform the Laboratory Manager in sufficient time to enable ordering and delivery prior to stocks running low.
- The Section Head or Senior Technician (Specialized analyses) is responsible for specific equipment and methods, especially trouble-shooting, maintenance and solving problems, as well as continuing training of junior staff when required. Training records for staff should be regularly maintained.
- Junior technician(s) are responsible for performing analytical work following Standard Operating Procedures (SOPs), under the direction of the Section Head or Senior Technician.

Note: Depending on the workload, one senior technician could be responsible for both Proximate and Specialized analyses.

This organization reflects an 'ideal' situation in a mature laboratory. In a new laboratory, however, the structure may initially be quite different. From the outset, the laboratory should have all expertise needed to perform all the methods it offers. In practice this means recruiting senior technicians with the background needed for the methods (see Section 3.2.1). This group should be regarded as the backbone of the laboratory, that trains additional personnel in case of an increasing volume of work and as a back-up for each assay. Ideally, over time, the structure should become similar to that discussed above.

If the laboratory is part of a larger organization, such as a feed producer or a research organization, its position and relationship with the other units should also be clearly described within its structure. The laboratory has a responsibility to ensure that the quality and credibility of its results are of the utmost quality. In an accredited laboratory this would require an independent QA Manager.

3.3 REALIZATION OF THE LABORATORY – PROCEDURES

Besides the physical completion of the laboratory, procedures such as SOPs, quality control programme, participation in proficiency (both internal and external) programmes and use of reference materials need to be put in place to ensure the laboratory consistently produces results to a high standard. These procedures should guarantee that all aspects of the analytical process are performed efficiently and are traceable. This should be documented in a set of SOPs that form the basis of the Quality Management System (QMS).

This section briefly describes the QMS procedures. The topic is considered in greater detail in Chapter 4.

Procedures should ensure that the laboratory can prioritize and organize its workload and guarantee the quality of results produced. The ISO/IEC 17025:2005 standard can be very helpful to identify which procedures should be prioritized. In the initial phase, however,

the focus should be on aspects that directly influence the quality of the results. These procedures are:
- Acceptance criteria for samples.
- Sample preparation.
- Description of methods, including validation of results.
- Quality control (first line of control).
- Maintenance and calibration records of equipment.
- Job descriptions, including responsibilities and continuing competence of individual technical staff.
- Training records of technical staff, covering which methods they can perform, level of training, whether they can perform a method independently or under supervision, and their ability to train others, etc.
- Traceability and storage of raw data.
- Cleaning procedures for the laboratory.

The presence and implementation of these procedures from the initiation of the laboratory will positively affect the quality of the results, and can be used as a starting point for the implementation of a comprehensive quality system.

3.4 CONTINUITY AND IMPROVEMENT OF THE LABORATORY

Success of the new laboratory depends strongly on its ability to respond to new opportunities and challenges in the commercial market or within the research field. For this purpose, the laboratory should focus on efficiency, expertise and innovation.

Efficiency is related to the volume of analytical work performed within a specific time interval. Increased efficiency leads to a lower unit price, making the laboratory more attractive for customers. Automation, such as the use of sophisticated equipment and autosamplers instead of manual methods, can be useful tools for increasing efficiency and sample throughput. Automation of support processes, such as recording of sample information, raw data and results, creating working lists, production of reports, etc., can strongly increase the efficiency of the laboratory by saving staff labour time and increasing the volume of work processed. This efficiency gain can be realized with the implementation and use of a specially designed Laboratory Information Management Systems (LIMS).

The laboratory should be constantly striving to improve efficiency and work throughput to ensure its continued success.

The second important factor for success is to provide a consistent high standard of service. This will give customers confidence and ensure repeat business. Laboratory expertise is strongly related to the specific knowledge, skill and experience of the technicians, which is often described as human capital, and this is directly linked to the quality of the laboratory. The laboratory should develop a policy to develop and guarantee the continuity of this knowledge by utilizing training programmes to ensure there is always a 'back-up' technician for every assay. It is essential that the laboratory make every effort to retain its experienced staff by creating a positive work environment, with good working conditions, fair rate of salary, clear career path, training opportunities, etc. This will promote accumulation of expertise, and consequently increase both the quality and also the market position of the laboratory. The importance of human capital should not be underestimated.

The third important issue is innovation, or research and development (R&D) to improve the quality and to broaden the scope of analytical possibilities. The first two issues mentioned, i.e. efficiency and expertise, are the basic conditions needed to build a successful R&D policy, which will enable the laboratory to respond successfully to new opportunities in the market. In general, a new method will only be developed and validated following a definite request from a customer for a large number of samples, probably ≥ 100. The laboratory must weigh the risks of developing new methods without any guaranteed samples against developing new methods for specific requests. To reduce risk, the laboratory manager should keep themselves up-to-date with current market trends by attending relevant conferences and seminars, as well as monitoring the latest published literature. In this way, being a market leader for new analyses will give the laboratory a clear advantage over competitors.

The key to a successful and long-lasting relationship between the laboratory and its customers is confidence. Customers must have full confidence in the quality of the data produced. The presence of a QMS, preferably accredited, enhances trust of the customers. Another important issue in the relationship with the customer is to meet the required deadlines and agreed prices (quotes should be given to prospective clients). To evaluate and improve the relationship, the laboratory should have regular contact with customers and engage in customer surveys to identify opportunities for improvement. This is an essential part of a QMS (see Chapter 4).

Procedures need to be developed to deal with complaints in a professional manner, and subsequently to solve them in an efficient and diplomatic way to ensure customer satisfaction with the overall laboratory service.

The level of communication, and also of additional services offered, gives an opportunity for a laboratory to stand out from its competitors. For example, a consultation service for nutritional information for various animal feed requirements, or an explanation of results, and personalized reports that might be on a fresh weight or dry matter basis as determined by the client's specific requirements, can sometimes tip the balance in favour of the laboratory.

Chapter 4
Implementation of a Quality Management System and the road to accreditation

4.1 INTRODUCTION

Right from the start, the focus of the laboratory should be the implementation of rules and procedures to guarantee the quality of the results produced. A QMS is the total set of rules and procedures that enables the laboratory to assure its quality. For this purpose, the system should cover each aspect of the laboratory that could influence either directly or indirectly the integrity of the analytical results. The range of these aspects makes the implementation complex, especially when starting a laboratory, and therefore international standards, such as ISO/IEC 17025:2005, have been developed to describe all areas that should be covered within the QMS. Full implementation of these standards ensures that all relevant aspects are addressed and that the QMS can be accredited according to an internationally recognized standard.

The first step in the process of implementing a QMS is to understand its principles and the correct interpretation of the standard. This knowledge is necessary as the standards only describe which aspects should be covered in the quality system, not how. Understanding the principles enables the laboratory to translate the requirements of the standard into procedures and rules. The next section (4.2) describes these principles and gives examples of how the different issues can be organized within the laboratory. The subsequent section (4.3) deals with the content of ISO 17025 in detail by discussing all aspects described and connecting them to the issues in Section 4.2. A road map for the total implementation of a QMS is the subject of the final section, 4.4. The sequence of implementation not only reflects the relative importance of the various issues, but also the requirements to prove that the system is operational and approved, which is essential for accreditation.

4.2 BASIC PRINCIPLES OF QUALITY

Aspects that affect the quality in the laboratory can be categorized on three different levels. Firstly, the technical level contains all aspects that directly influence the quality of the analytical process. Secondly, the organization level contains all relevant aspects within the organization itself that indirectly influence the quality of the analytical process. Finally, the commercial level contains all relevant aspects of the interaction between the laboratory and its customers. These three levels will be described in detail in the following sections.

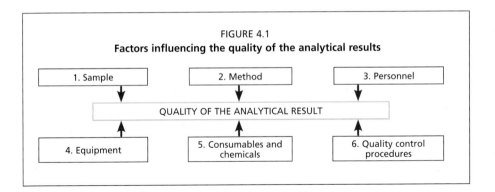

FIGURE 4.1
Factors influencing the quality of the analytical results

4.2.1 Technical level
The technical level contains elements that directly influence the quality of the analytical results produced (Figure 4.1), and the six parameters discussed in the subsequent sections.

4.2.1.1 Sample
The physical condition of the sample material should allow a reliable performance of all determinations requested. For this purpose, the laboratory should set up procedure(s) to establish the acceptance criteria and preparation of individual sample types. For acceptance, the physical state of the sample's status on arrival should be noted (i.e. temperature, frozen, partially thawed, etc.), and also the form in which the sample needs to be for each assay, whether freeze dried, defatted, fresh, etc. Other factors, such as the minimum amount of material required for each test, should be listed and made available to customers (see Section 4.2.3). The laboratory should ensure that the test portion used for the determination is representative of the total sample provided. For animal feeds this is generally achieved by prior drying (oven or freeze drying) and grinding, leading to fine, homogeneous material that allows a representative sample to be taken. Products with a high fat content may need a different type of grinder (not forced through a mesh) or extraction of the fat prior to grinding. International standards for general sample preparation are available (ISO 6498 Animal feeding stuff – Preparation of test samples), with additional requirements described in standards for specific determinations.

It is the laboratory's responsibility to prove that their sample preparation procedure and storage lead to reliable results within acceptable variation limits. This proof can be part of the validation studies for specific determinations (see 4.2.1.2 Methods, below).

4.2.1.2 Methods
The laboratory should base its selection of a specific method on the requirements of the customer, the technical capabilities of the laboratory (see Chapter 3), the availability of verified and referenced methods (ISO, AOAC International, etc.) and, very importantly, sample type. In the case of animal feedstuffs, the methods are often based on international standards (see Section 3.2.1), which means the laboratory does not have to prove the correctness of its methodological principle. If the laboratory uses an in-house developed method, or modification of an accepted, verified method, it must validate the method to demonstrate that it is 'fit for purpose'. To validate the method, a robust SOP must be written and a validation

study performed. The validation study must demonstrate the robust nature of the method and consistency of results obtained using verified reference material.

The analytical protocol (SOP) describes exactly how to conduct the method and contains all vital information about its determination. This information should emphasize all critical steps, which reflects the knowledge and expertise within the laboratory, and data about the quality of the determination. To demonstrate the traceability and repeatability of the analytical results, it is critical that the document reflects actual practice, using the equipment and resources available in the laboratory. The format of the SOP should be comparable to that of international standards, dividing the SOP into different sections that describe principles, scope, limit of detection, repeatability, chemicals, equipment, procedure, calculation and interpretation of results. The laboratory should have procedures to enable the creation and maintenance of SOPs (see Section 4.2.2, document control). The SOP should contain information relating to any health and safety issues and environmental considerations as appropriate.

The general aim of a validation study is to prove that the principle of the method is correct and the quality of results is based on accuracy and precision. The protocol of a validation study, however, varies between methods and depends on its purpose and scope, so understanding the principle of the method is vital. Validation is achieved by determining the following parameters delineating the quality of the method:

- **Limit of detection** (LOD) and **Limit of quantification** (LOQ) are the lowest concentration that can be identified and measured, respectively.
- **Accuracy** describes the difference between the result found by the laboratory and the true value in a sample.
- **Precision** describes the variation in the results found by the laboratory in the same sample. Precision can be estimated at the same time and under the same conditions, which is repeatability, and at different times and conditions, which is intra-laboratory reproducibility.
- **Linearity** describes the upper limit for which the concentration shows a linear relationship with the measured signal. This parameter is only relevant with a calibration curve.
- **Selectivity** describes the influence of other components on the measured signal.
- **Sensitivity** describes the quantitative relationship between the analyte and the measured signal.
- **Robustness** describes the effect of variation in the procedure on the measured signal.
- **Stability** describes the change of concentration of an analyte over time, stored at specific conditions, such as temperature and pressure.

Calculation of these parameters requires analytical and statistical knowledge, and depends on the type of methodology utilized. In Appendix A, examples of these calculations are given for different types of methods.

A validation study does not have to include all of the above parameters (see Appendix C). The determination of selectivity and robustness is generally limited to a newly developed in-house method or if modifying a standard method, whereas the determination of linearity is only relevant for spectroscopy and chromatographic determinations. LOD, accuracy and precision should always be part of a validation study. In general, the laboratory should estimate these parameters and record these values in its analytical protocol. If the laboratory, however, claims to work according to a standard method, it should prove that its values are at least comparable to those stated in the official method. The analytical protocol should contain the values mentioned in the official method.

Although there can be some interaction between the performance of the validation study and the analytical protocol, it is important to emphasize that the determination of parameters such as LOD, accuracy, and precision are based on the final analytical protocol. If changes are made to the protocol, the laboratory should investigate the effects on these parameters by conducting an additional validation study. Results for the validation study should be available on request and should be the basis for calculating the measurement of uncertainty (see Appendix D).

4.2.1.3 Personnel

Knowledge and skills of laboratory technical staff affect the quality of the results produced. To guarantee this aspect, the laboratory should be confident that the personnel involved are capable of conducting the methods correctly. The laboratory should have a procedure for training and authorization of their personnel for each determination. This should be described in the training and authorization records for each technician (see Appendix G). This document could contain a check-list with the information required to conduct the method and on which results the authorization is based. Only staff with training authority can train others.

An authorization matrix, as given in Table 4.1, is a useful tool to provide an overview of training status. From the example it can be seen that only one person is authorized to conduct the analysis 'method D'. Ideally, there should be at least two technical staff capable of conducting each of the laboratory's standard methods.

Temporary personnel should also be trained and authorized for each determination. Training records can be divided into stages of the method, e.g. digestion only or spectrometry, as well as status of training level, e.g. proficient, or capable under supervision.

On-going competency must also be demonstrated in training files. This may use participation in external proficiency schemes, 'Ring Trials', inter-laboratory comparisons, etc. Should a member of staff fall below the requirements of on-going competency they must stop performing the analysis until competency is regained.

TABLE 4.1
An example authorization matrix

Technical staff member	Determination			
	A	B	C	D
1	✓	✓		
2		✓	✓	
3	✓	✓	✓	
4				✓

Notes: ✓ = authorized to perform

4.2.1.4 Equipment

The equipment affects the quality of the analytical results produced. In general, the specifications of more sophisticated equipment improve the quality parameters, such as a lower LOD and better precision. These effects are described in the analytical protocol and the laboratory should ensure that equipment is working according to these specifications. For this purpose the laboratory should focus on regular maintenance and performance checks.

The aim of maintenance is to avoid future problems with equipment, which could lead to unreliable results. Maintenance of equipment can be divided into regular checks done by laboratory technical staff, and more sophisticated servicing performed by external specialist contractors. The frequency of both types of maintenance depends on the type of equipment. For standard equipment, maintenance by laboratory technical staff is generally sufficient, with external contractors necessary only in the event of a major malfunction. For more sophisticated equipment, such as chromatography equipment, high-speed centrifuges, etc., regular servicing by accredited staff from the supplier or a specialist contractor is essential to guarantee the continued performance of the apparatus.

Performance checks are conducted to prove that the equipment is meeting the required specifications. Calibration of volumetric equipment, control of balances, and estimation of the wavelength in a spectrophotometer, are examples of these kinds of checks, which should be described in protocols containing information about the procedure, frequency of checks, and criteria. The performance check procedures can be divided into method-dependent and method-independent procedures. Method-independent checks of equipment are undertaken without conducting a specific analytical method, but rather by measuring general physical properties such as weight, temperature and volume that can be traced back to internationally accepted references. The checks for most analytical equipment, however, can only be undertaken by conducting a method, and should focus on issues such as sensitivity (minimum response for a calibration solution) or retention behaviour (retention time for a specific compound) in the case of chromatographic assays.

For all main pieces of equipment, the laboratory should have two documents: an operational manual (or User Guide) and a logbook. The operational manual will contain all relevant information (such as protocols and frequency) about maintenance and performance checks, and information on the safe use and operation of the item of equipment. The logbook is used to record all maintenance, problems and results of checks (oil change, breakdowns, etc.) performed for the apparatus involved, and also to record the extent of use of the equipment and names of the users. The presence of these documents and a properly filled-out logbook demonstrates that the equipment is working according to the specifications and is capable of producing reliable results. All critical equipment should be labelled with a unique code, and a spreadsheet or list should be used to ensure that all maintenance and checks are performed according to a fixed schedule.

4.2.1.5 Consumables and chemicals

Impurities in and contamination of consumables and chemicals can negatively affect the quality of a determination, and therefore the laboratory must ensure their quality. Critical chemicals and consumables should be explicitly described in the analytical protocol of the determination, including criteria such as 'HPLC-grade only', 'containing <0.01 ppm lead', etc. Certificates for purity should be obtained from suppliers. If using commercially available stock solutions, such as for minerals and amino acids, the laboratory should verify that suppliers of the reference materials are ISO 34:2009 certified. If not, critical levels should be checked against an independent reference standard.

4.2.1.6 Control of the results

The previous aspects focused on the creation of optimal conditions to produce reliable results. The day-to-day routine, however, is also influenced by random variation introduced during execution of a procedure, introducing an error. Therefore the laboratory should constantly check the quality of the results produced by the implementation of a first line of control procedure. The basic principle is to analyse a control standard within each batch of samples and use its result to judge the quality of the other results in the batch. The matrix of the control standard should be representative of the samples generally analysed by the laboratory. The most important demands for these standards are stability and homogeneity, which should be guaranteed by the laboratory. The laboratory should have criteria to judge the results of the control standard and a method to record these values. In practice, Shewhart-charts are most commonly used to record these results, including acceptance criteria and relevant statistical information (see Appendix B). Shewhart-charts are very powerful tools for a continuous check on the quality of the results, and also reflect changes in the performance of the method over time. The first line of control, however, only recognizes problems at batch level and is therefore not guaranteed to avoid errors in individual samples. For this purpose, it is advisable to conduct analyses in duplicate in the same (repeatable conditions) or different batches (intra-laboratory reproducible conditions). Analysing samples in different batches is statistically the best, but not always the most efficient approach. Quality of the results should also be monitored using set criteria for error percentages allowable between duplicates for each sample type.

Participation in collaborative trials (proficiency testing) and the analysis of Certified Reference Material (CRM) give valuable information about the accuracy of the results of the laboratory. The results found in collaborative trials are mostly evaluated by a *z*-score, which is calculated as the difference between the found value and the average divided by the overall standard deviation, and should be between -2 and +2. The *z*-score should be used to identify the presence of systematic errors, and the determination of the uncertainty of measurement. For the evaluation of the *z*-scores, the laboratory should be aware of the effects of differences among the methods used by other participants.

Two points need to be noted when referring to the estimation of accuracy when using the results obtained from collaborative trials or from analysing a CRM. Firstly, the average obtained for a collaborative trial is based on a limited number of laboratories and methods; it should therefore be regarded as a compromise instead of the true value. The advantage of certificated materials is that its results are generally based on a range of methods, which decreases the influence of systematic errors and therefore can more closely represent the true value. Secondly, the results obtained for most proximate analyses are empirically connected to a specific method. An example is crude protein, which is for most animal feedstuffs by definition $N \times 6.25$, whereas the true protein in a sample can systematically deviate

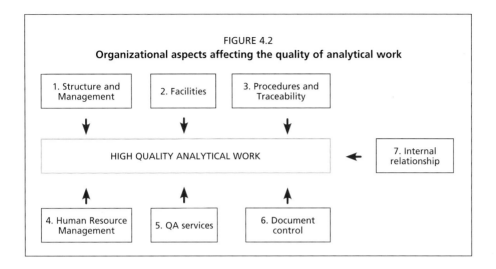

FIGURE 4.2
Organizational aspects affecting the quality of analytical work

from this value. It is important to realize that, for these determinations, accuracy is based more on convention than on the true chemical composition.

4.2.2 Organization level

The analytical work is fundamental within an organization to facilitate its conduct and continuity. Different aspects of the organization are linked to the analytical process and therefore indirectly influence the quality of results produced (see Figure 4.2). These are more general aspects for organizations rather than being laboratory specific, and can therefore also be found in ISO 9001:2008 Quality Management Systems – Requirements. If the laboratory is part of a larger organization, it might use existing procedures for these aspects, otherwise it is required to develop and implement these issues by itself (see Section 4.4).

These aspects are described below in more detail.

4.2.2.1 Structure and management

Work in a laboratory relies on teamwork, which should be organized in a manner suitable to avoid the production of incorrect results. A description of responsibilities relating to the validation of methods and authorization of results should be clearly described and conducted only by skilled laboratory personnel. To guarantee the continuity of an analytical process, the head of each section should nominate a deputy that can take responsibility for the work in their absence.

The QA Manager in the laboratory has a special position because of their responsibility for monitoring operational quality within the laboratory. This person should have an independent position, with direct access to the upper management of the organization. Sometimes the QA Manager may be part of a separate QA unit (see Section 4.2.2.5 below) to guarantee independence.

4.2.2.2 Facilities

Basic and specific facilities are needed to conduct laboratory work and should be located at a permanent location, as described earlier, in Chapter 3. The laboratory is obliged to prove that the status of these facilities does not negatively affect the quality of its work by isolating any interfering activities (such as keeping sample grinding separate from analytical work) and by following a daily cleaning programme to minimize contamination or sample errors.

Specialized equipment will generally require service and maintenance from external services. The facilities that house these items of equipment should be controlled by the laboratory to ensure continuity of operation, such as required temperature, uninterrupted power supply, dust free environment, etc.

4.2.2.3 Traceability and procedures

Traceability of all activities in the laboratory is necessary to guarantee the quality of each individual result. For traceability, it is essential that steps within an analytical process are clearly described in the appropriate SOP, and that their correct performance can be demonstrated (traceability).

The analytical process should be described in a set of documents that covers general and more detailed procedures, as described in Section 3.3. These procedures should be as explicit as possible to avoid variation in performance.

To prove correct performance, all vital information must be recorded and stored. An example of this is a sample's worksheet, which as a minimum should include:
- Date of analysis
- Condition of sample on arrival (frozen, fresh, dry (oven dried or freeze-dried), etc.)
- Information about the control samples analysed
- All raw data produced, including spectrometry printouts, weighing sheets, etc.
- Name of laboratory technician for each analysis conducted
- Identification of equipment used (e.g. ID number of balance, centrifuge, etc.)
- Signature for acceptance and date the report is sent to client
- Any communication from the client

All critical factors should be traceable from the information on the worksheet. If necessary, the laboratory can prove that the technician was authorized to conduct this determination, the equipment had the correct calibration status, calculation of results was correct, samples were connected to the first line of control, and the results were authorized by qualified personnel. It is essential that all of this information is available and stored correctly for a specified period by the laboratory. The storage time for information depends on legal and customer requirements, but, in general, a storage period of 5 years is acceptable.

At times procedures may deviate from the standard protocol. This may be at the client's request or due to insufficient sample amount. The laboratory should implement a non-standard work procedure that gives flexibility to deal with this kind of situation. The key elements of this procedure are authorization and recording that a non-standard method was used due to client specifications, with a record of any deviations from the standard method. The client must be notified personally, and additionally in the laboratory report it may be necessary to disclaim any accreditation for the test concerned. In general, the laboratory should have a restrictive policy on such issues, and their decision about acceptance should be mainly based on the effect of the deviation on the quality of the results. In the case of non-standard work, traceability will become an even more important issue.

4.2.2.4 Human resource management

The knowledge and skill of the laboratory personnel strongly affects the quality of the laboratory's results, as mentioned in Section 3.4. The technical skills of the personnel should be recorded in the training and authorization records as previously mentioned. At this level, the focus is more on general aspects of this policy, such as the development and motivation of personnel. The aim of this policy should be to guarantee the continuity of the laboratory and encourage continual improvement.

In larger organizations, these activities are concentrated in a special unit that deals with personnel affairs in general and is also responsible for archiving all relevant documents, such as training records. For stand-alone laboratories, the management is generally responsible for this aspect.

4.2.2.5 Quality Assurance services

The QA service plays a vital role in the process of implementation and maintenance of the quality system within organizations and laboratories. In a larger organization, these activities can be organized in a separate unit. To guarantee its independent position, direct access to upper management should be a priority. The first task of this unit is the development of a quality policy and the implementation of the basic principles of quality within the organization, mostly achieved by the implementation of ISO 9001:2008. Additionally, this unit should also facilitate the development and implementation of specific procedures within individual units. In the case of laboratories, the QA unit should help with the interpretation and implementation of ISO/IEC 17025:2005, leading to setting up of laboratory SOPs. The second task of the QA unit is to check the operational quality system and facilitate its continual improvement by conducting internal audits that check how the system works in practice, and identifying where improvements could be made. The unit can also perform external audits of product suppliers, sub-contractors or services used by the organization. The great advantage of centralization of QA activities is accumulation of knowledge and skill in this unit, which is beneficial for the whole organization as well as training internal auditors. Success of the quality policy, however, also depends on the relationship between the QA unit and the laboratory itself.

A separate QA unit is usually absent in most small stand-alone laboratories. In this situation, these activities should be conducted by the QA Manager in the laboratory itself or employ external expertise to create and implement the basic QMS, including conducting internal audits.

4.2.2.6 Document control

The quality control system requires a great number of controlled official documents to describe all aspects of the process. These documents have to be updated regularly and the organization must ensure that only the most recent version is available and in use by personnel. For this purpose, the organization should have a procedure for document control that regulates administration, distribution and revision, to ensure that only the correct versions of documents are in use.

In larger organizations, document control is mostly centralized and conducted by the QA unit. Sophisticated electronic systems have been specially developed to facilitate the total process of document control. In the case of smaller laboratories, document control can also be regulated by the use of a database system.

4.2.2.7 Internal relationships
If part of a larger organization, the laboratory will have acquaintances and interactions with other internal units, which could affect the quality of results. An example of a misuse of this interaction is an internal unit giving short deadlines to the laboratory to complete analyses. The organization has the responsibility to identify such influencing factors and ensure that they do not negatively affect quality. An organizational structure that ensures the laboratory has an independent position can help to avoid these issues.

Another example is the relationship between the laboratory and upper management. The implementation and maintenance of a QMS involves various investments that should be approved by upper management, otherwise it will risk functioning under sub-optimal conditions. To guarantee that the laboratory gets the resources required, commitment from upper management is essential.

4.2.3 Commercial level
The commercial level is the relationship between the laboratory and customers, which can affect the quality of analytical results. This subject is a relatively new element in QMS and its focus is on the perceived value of the result for the customer as a decision-maker. This value depends not only on technical qualifications but also on aspects such as time and price. Customers have expectations on these issues and therefore laboratories are obliged to be realistic and ensure the customer is aware of constraints such as turn-around time, price and methodology.

These aspects will be described based on the three phases of communication between the laboratory and the customer.

4.2.3.1 Acquisition phase
The acquisition phase starts with the first contact between both parties, and ends with the acceptance of the samples by the laboratory. The aim of this phase is to define the customer's expectations and laboratory services required. General information about the laboratory services, such as methods, quality, prices, criteria for acceptance, turn-around time and general conditions should be freely accessible for customers (for example on a Web site or as part of a quotation). If standard laboratory services are required, this information will generally be sufficient for the customer to make an informed decision.

It is advisable that laboratories have their own submission document that can be used by customers, including acceptance criteria for standard services, and test options. The use of this document can avoid confusion and disappointment regarding expectations of the services available. This submission document also forms a contract between the client and laboratory.

For special work outside the routine, it is advisable to have direct contact between both parties that should lead to a contract that specifies the work and the expectations (methods used, delivery time and price). Outsourcing of tests should also be explicitly described in this contract. These contracts should be signed by both parties and stored by the laboratory.

4.2.3.2 Operational phase

The operational phase starts after the acceptance of the samples, and ends with sending the final report to the customer. In the case of problems or deviations from the standard procedures, the laboratory is obliged to inform the customer and to discuss the significance of the problem and other options available. These deviations from the standard protocol should be treated as a non-standard procedure and conducted accordingly (see traceability, Section 4.2.2.3). In this situation the laboratory remains responsible for the quality of the results and therefore should make a judgement between its commercial interests and the quality of the results. The laboratory should retain all correspondence with the customer regarding these issues.

4.2.3.3 After-sales phase

This phase starts when the customer receives the final report. The report should give a clear overview of all essential information about the sample, such as the identification code, sample description, type of analyses conducted and the results, including the relevant units. Accredited laboratories should also note which determinations were conducted inside the scope of the accreditation and which were conducted outside the scope of accreditation. Any tests that were sub-contracted should also be clearly indicated.

The laboratory should deal with customer's queries about the results and interpretation in a professional manner. Communication with customers should be limited to laboratory management to ensure a consistent policy in answering questions. Direct communication between customers and technical staff should be avoided.

On completion of the analytical work, storage of data and sample material should be organized. General standards for the storage of samples (see technical level, sample – Section 4.2.1.1) and of data (see organization level, traceability – Section 4.2.2.3) should be documented. Customers should only have access to data from their own samples to guarantee confidentiality. After a fixed interval (e.g. 1 calendar month post-reporting), samples should be destroyed or disposed of in a suitable manner.

Two additional important issues in the communication between the laboratory and the client are complaints and feedback. The laboratory is obliged to have a procedure for dealing with customer complaints. These complaints are useful to identify opportunities for improvement and to help improve the quality and level of service to the customer. Taking complaints seriously will also increase customer confidence. The laboratory has a responsibility to seek feedback from customers regarding its performance. A questionnaire is a good tool to obtain this information, and is also useful to optimize processes in the laboratory.

4.3 READING AND INTERPRETATION OF ISO/IEC 17025:2005

This section describes the different aspects mentioned in the standard and makes a connection with the general principle of the quality system as described in Section 4.2. This information can help with understanding and therefore assist with implementation of the different elements of the standard.

The most important sections of the standard are in its Section 4, Management requirements, dealing with general demands regarding the organization and quality system at the organization level, and in its Section 5, Technical requirements, which is more specific for laboratories. In Boxes 1 and 2 the present authors focus on important items in ISO/IEC

BOX 1
Important sections from Section 4 of Standard ISO/IEC 17025:2005

4. Management requirements

4.1 Organization

This addresses the Organization level, its structure and management.

General requirements are stated, such as legal status, extent of management system responsibility, capability to perform testing activities according to the ISO Standard, quality system and customer requirements, as well as the provision of adequate facilities. Special attention is paid to the problem of improper influence due to conflict of interests, especially where the laboratory is part of a larger organization.

The laboratory should have an independent position and its structure, responsibility, authority and interrelationships should be fully defined. Personnel should be familiar with the procedures and the relevance of the quality system. The laboratory should have a quality manager and appoint deputies for key managerial personnel.

4.2 Management system

This addresses QA services and internal relationships at the Organization level.

The laboratory should have an implemented management system that covers all relevant aspects to guarantee the quality of the services offered. An overall Quality Manual should give an overview of all aspects of the management system and refer to more detailed information, with an outline of the structure of the documentation. Upper management should play an active role in the creation, improvement and maintenance of the quality system and should provide evidence of commitment, communicating its importance and ensuring integrity if changes are made. This active involvement should prove that quality is an essential part of company policy and that the resources required will be made available.

This commitment should be in the form of a written, signed document that contains a statement about the importance of the quality system for the organization as a whole.

4.3 Document control

This addresses document control at the Organization level other than control of data related to testing and calibration, which is covered elsewhere.

Procedures are required to control all documents, in the broadest sense, that are part of the management system, including manuals and software. All documents produced should be approved by authorized personnel, and a master list should identify the current version and distribution to preclude the use of invalid or obsolete procedures. The documents should be uniquely coded and periodically updated. Changes in documents should be marked in the updated version and be approved by authorized personnel. The standard offers the opportunity to make minor changes by hand on the condition that they are signed and dated and a revised document is re-issued as soon as is practicable.

4.4 Review of requests, tenders and contracts

This addresses QA services at the Organization level.

The laboratory should administer requests, tenders and contracts. This also covers work that is subcontracted to other laboratories. The standard however distinguishes between repetitive routine work and more complex and advanced tasks. In the former case, the laboratory can refer to a general agreement with the customer. The standard also demands communication with the customer if deviations or amendments are required.

4.5 Subcontracting of tasks
This addresses the Commercial level at the acquisition phase.

For subcontracting, the work should be placed with a competent subcontractor that works according to the ISO standard. The laboratory remains responsible for this work unless the customer specifically requests a particular subcontractor. The laboratory should have a list of suitable subcontractors who meet their criteria.

The use of subcontractors is a delicate issue because the laboratory remains responsible. Good knowledge of the quality of other laboratories is essential to avoid subsequent problems. In the absence of this knowledge, it is prudent to avoid subcontracting, and rather allow customers to deal directly with other laboratories.

4.6 Purchasing services and supplies
This addresses QA services at the Organization level.

A policy should be available for purchasing services and supplies that can affect the quality of the tests (pH buffers, analytical standards, etc.). These products should be stored in the correct manner, and before use checked against the defined requirements. Records regarding this testing should be kept. A list and an evaluation of suppliers of critical consumables should be available.

4.7 Service to the customer
This addresses the Commercial level in the after-sales phase.

The laboratory should cooperate with customers to clarify any requests regarding laboratory performance.

This transparency must not interfere with the confidentiality of other customers. Therefore customers should only be able to access their own results and not those of other clients. In the case of an inspection, the laboratory should only make available relevant reports and results.

The laboratory should seek feedback from its customers that can be used to improve the management system.

4.8 Complaints
This addresses the Commercial level in the after-sales phase.

A policy and procedure for the resolution of customer complaints should be available. Records about the complaints and the resulting investigations and corrective actions should be kept.

4.9 Control of nonconforming testing and/or calibration work
This addresses the Organization level regarding traceability and procedures, and also applies to the operational phase of the Commercial level.

The laboratory should have a procedure to be followed if a deviation from its own procedures or from the original agreement with the customer is required. *Examples of these deviations are technical problems within an assay, which might lead to less accuracy, or where internal procedures have not been implemented.* To deal with these situations correctly, the Standard demands that this procedure should contain the following elements:
- Responsibility and authority for the management of non-conforming work and how to identify these, as well as the response actions required.
- Evaluation of the significance of the changes (impact assessment).

(count.)

- Corrections should be made, and customer notified.
- Responsibility for authorizing the resumption of work should be defined.

If a non-conformity leads to doubts about the quality, corrective actions should be undertaken (see 4.11).

4.10 Improvement
This addresses QA services at the Organization level.

The laboratory has an obligation to continuously improve the effectiveness of its management system through the use of quality policy, audit results and preventive and corrective actions. *The development of a quality system is a dynamic process that should constantly be evaluated, developed and improved. The demands of the quality of the system will increase in time, which gives the new laboratory the opportunity to develop their system within a certain time period.*

4.11 Corrective actions
This addresses QA services at the Organization level.

Corrective actions should be conducted for all non-conforming tests or if a departure from the management system or technical operation has been identified (see 4.9). A procedure for corrective actions should contain the following information:
- Determination of the cause of the problem.
- Identification of potential corrective actions.
- Determining the efficiency of the corrective actions.
- Additional audit (see 4.14) to evaluate the new situation.

Date and initials of the person(s) initiating and following up on the corrective action.

The whole process should be documented, including all results obtained by the corrective actions. *This procedure should also be followed when dealing with deviations found during internal and external audits. It should focus on the impact of the problem, looking for a solution and to guarantee that the problem is solved.*

4.12 Preventive action
This addresses QA services at the Organization level.

A policy is needed to identify potential sources of nonconformities in the management system or technical areas. After identification, improvements or preventive actions are required that should be conducted according to a procedure, this includes the initiation and application of controls to ensure they are effective. *An example of a preventive action is the replacement of poorly functioning old equipment, which avoids the issue of future tests being out of specification.*

4.13 Control of records
This addresses methods at the Technical level and procedures and traceability at the Organization level, both in general and for technical records.

All relevant quality documents should be uniquely identified and stored in an appropriate way by the laboratory to avoid damage and unauthorized access or destruction. Records should be held securely and in confidence (see 4.7). Technical records should contain all relevant information to identify the factors, including personnel involved, that could affect uncertainty and enables measurements to be repeated if necessary. The records should be stored for a specified period. Changes in records must be traceable and signed by the person making them.

The presence and access to this information is essential for traceability and control of the quality of the tests conducted by the laboratory, giving customers and auditors confidence in the QMS.

4.14 Internal audits

This addresses QA services at the Organization level.

The standard demands the presence of a schedule of audits that covers **all** elements of the QMS and that should be conducted periodically. The QA Manager is responsible for the organization of these audits, which should be performed by trained and qualified personnel that have, if possible, a position independent from the activities being audited. *The Standard allows small, stand-alone laboratories to use its internal personnel to conduct these audits, but auditors must not audit their own work.* If the audit reveals doubts about the effectiveness or validity of a test it should immediately be followed up with a corrective action (see 4.11).

Records should be kept covering the area of activity audited, together with the findings and corrective actions. Follow-up audits should verify and record the effectiveness of any corrective actions required.

Although the audits should cover all aspects, the laboratory should set priorities for items and establish the frequency for performing these. In the beginning, technical audits have a higher priority and should be performed more frequently than general audits. For smaller laboratories, it could be wise to hire external expertise to conduct specific audits and train internal personnel. The laboratory has an obligation to improve the quality of audit records over time.

4.15 Management reviews

This addresses internal relationships at the Organization level.

The laboratory management should periodically undertake a review of their QMS to ensure its effectiveness and introduce changes or improvements as needed. This review should address aspects such as: suitability of the existing policy and personnel; internal audits; corrective and preventive actions; assessment conducted by external auditors or accreditation bodies; results from proficiency tests; changes to the type or amount of work; customer feedback; recommendations for improvement; and other relevant factors. Activities for any such review should be performed within an agreed timeframe.

BOX 2
Important sections from Section 5 of Standard ISO/IEC 17025:2005

5. Technical requirements
5.1 General
The laboratory should have the resources and policies in place enabling it to perform the tests in compliance with the stated methods and therefore guarantee the quality of the results produced.

5.2 Personnel
This addresses the Technical level for personnel and the Organization level for Human Resources Management.

The laboratory should ensure the competency of its personnel to deal with specific equipment, perform tests and sign reports. This competency should be based on education, experience and skills. Supervision should be provided during training. Management should set up a programme for the education, training and continuing competency of personnel, and programme effectiveness should be evaluated. In the case of temporary personnel, supervision should be ensured by the laboratory. Job descriptions should be available for all staff positions. Laboratory management is responsible for the authorization of specific issues, which should be documented (including date of authorization) in the training files.

5.3 Accommodation and environmental conditions
This addresses facilities at the Organization level.

The facilities should allow tests to be performed correctly with adequate monitoring and control of any environmental conditions that could negatively affect the quality of the results. Incompatible activities should be separated by time or space to avoid cross-contamination, and access to all areas should be controlled. Good housekeeping, including appropriate cleaning and monitoring procedures, is needed to guarantee the quality of work and avoid contamination.

5.4 Test and calibration methods and method validation
This addresses methods at the Technical level.

The laboratory should use appropriate methods and procedures for all tests, including aspects such as sampling and storage, with relevant statistical information plus, if necessary, any estimation of the measurement of uncertainty. Instructions for the use of equipment and other relevant documents should be available for personnel at point of use. Any deviation to the method must be documented, technically justified, authorized, and accepted by the customer. *The standard allows the use of international or nationally recognized methodologies as protocols as long as they fully cover the procedure performed.*

The selection of the methods should be based on the demands of the customer. However, it is preferable to use national or internationally recognized standard methods. The laboratory should use the latest version unless there are serious constraints. Laboratory-developed methods (in-house methods) or modifications to recognized methods may be used if validated and the customer has been notified.

Validation of a method demonstrates that particular requirements for a specific intended use are fulfilled, and validations should also be performed for non-standard and laboratory-developed methods as well as standard methods used outside their scope, such as a different sample type. The scale of the validation depends on the needs of the application, but should always be a balance between cost and benefit. The range and accuracy of validated methods should be determined, and be relevant to customer needs. Data for validation studies should be stored. *The requirements for validation are described in section 4.2.1.2 of the main text and in Appendix C.*

The laboratory should have a procedure to estimate the uncertainty of measurement that takes into account all relevant factors (i.e. different technicians or operators, different equipment, etc.), using an approved statistical calculation. *This calculation gives information about the uncertainty in an individual measurement and is expressed as a 95%-confidence interval. This calculation takes into account random variation as well as systematic variation obtained for reference materials. This calculation is complex and the person conducting it should possess appropriate statistical knowledge. The laboratory can use ISO 5725:1994 for additional information about this issue. Appendix D gives some examples of this calculation.*

Calculations and data transfers within the method should be checked in an approved way. When using computers for this purpose, the laboratory should demonstrate that any software developed by the user is documented and validated, and that data is protected and the integrity and confidentiality of data transfer is guaranteed. *The use of standard software, such as spreadsheet programs, is allowed by the Standard.* Computers involved should be systematically maintained.

5.5 Equipment

This addresses equipment at the Technical level.

The laboratory shall have the equipment to enable correct performance of the tests to achieve the accuracy required. Calibration programmes should be installed for key quantities or values that should be checked before use. Each item of equipment should be uniquely identified and only be operated by authorized personnel. For each item of equipment that is significant to the test, a record should be made that contains at least the following information: identification; description of apparatus; checks of specification; location; manufacturer's instructions; calibration records; maintenance plans; and any damage. The laboratory should have procedures for safe handling and maintenance of equipment to prevent contamination or deterioration. Equipment that is outside specified limits should be taken out of service by removal or clearly labelling as 'out of use'; it should only be used after repairing and testing (as in Section 4.9 of the Standard). If possible, the equipment should bear a label that indicates the status and expiry date of calibration. If the equipment is temporarily placed outside the laboratory, the laboratory is responsible for checking its performance before being used again. Procedures should be available for intermediate checks. Special attention should be paid to the correct use of correction factors for equipment.

5.6 Measurement traceability

This addresses equipment at the Technical level.

All equipment that is critical should be calibrated before being put into service (see Standard section 5.5). Calibration of equipment should, if possible, be traceable to the international system of Units (SI). *For chemical laboratories, such calibration is limited to physical parameters such as weight, volume and temperature.* Otherwise the use of certified materials or participation in inter-laboratory comparisons should be implemented.

(count.)

Reference standards and materials should be calibrated by a body that can provide traceability. Checks should be performed to maintain confidence in the status of these reference standards, whilst their storage should be according to supplier recommendations to protect their integrity.

5.7 Sampling

This addresses the Technical level for the sample, and the Commercial level for the sample acquisition phase.

The laboratory requires a sampling plan and procedures based on statistical principles when it carries out sampling of substances. This plan should be available at the location of the sampling. *Sampling is a defined procedure whereby the part that is taken is representative of the whole, such as a batch of a diet. This sampling should not be confused with taking a test portion from a sample in the laboratory.* Deviations wanted by the customer should be recorded in detail and be included in all related documents, including test reports (see Standard section 5.10) and communications to personnel. All documents relating to the sampling strategy and performance should be archived.

In practice, sampling can lead to confusion and problems between customers and the laboratory. Two situations can be identified. Firstly, the laboratory receives a sample from the customer and conducts the required tests. In this case, the laboratory should state that the results are representative for the received sample material by using a disclaimer (see Standard section 5.10). Secondly, the customer asks the laboratory to perform a test from a larger batch, which includes setting up a sampling strategy. In this case, the laboratory is fully responsible and accountable for the sampling. It is important to discuss both options with customers.

5.8 Handling of test and calibration items

Addresses the Technical level for the sample, and the Organization level for traceability.

Procedures for transportation, handling, protection and storage of test items (i.e. samples) should protect their integrity and protect the interests of both the laboratory and the customer. *The laboratory should give each sample a unique code or number that is used in all test reports; these codes might be automatically generated if a LIMS system is in place.* Deviations from the standard protocol should be recorded in case there are doubts about the suitability of an item for a test or if it does not conform to the description provided, and the laboratory should contact the customer before proceeding and record the discussion and agreement. The laboratory should also have procedures and facilities to store and secure test items to guarantee their condition and integrity. *Storage conditions of laboratory samples depend on the type of sample and the tests required, and should be both part of the validation study and they should also be stated in the appropriate methods. For storing samples, a general procedure should be available. In general, the laboratory should have the capability to store at -20 °C, 2–8 °C and +20 °C, and in exceptional cases also at -80 °C.*

5.9 Assuring the quality of test and calibration results

This addresses the Technical level for quality control procedures.

The laboratory should have quality control procedures for monitoring the validity of tests undertaken. Data should be recorded in such a way that trends are detectable and results can be statistically evaluated. This monitoring should be planned and reviewed, and it might include regular use of certified reference materials, participation in external proficiency testing, replicate tests, re-calibrations and correlation with different characteristics. If quality control data are displayed graphically, trends are quickly noticed, and if outside the pre-defined criteria action should be taken to identify and correct the problem and prevent incorrect results being reported.

5.10 Reporting the results
This addresses the Commercial level in the operational phase.

Results should be reported accurately and must include all information requested by the customer and required for the interpretation of the tests. *The standard allows a simplified way of reporting if this is previously agreed with customers. T*he test report should contain at least:
- Title, name and address of the laboratory.
- Unique identification of the report, include numbering of pages.
- Customer.
- Identification of method.
- Identification of the item tested.
- Date the test item was received.
- Test results, including units.
- State of sample (as received; on dry-matter basis; fresh weight; etc.).
- Name and signature of person(s) authorizing the test report.

When required, a statement that the results relate to the items tested as received can be made (see Standard section 5.7). *Reports that cover all the above-mentioned items are generally sufficient for routine analyses. Results that are obtained from accredited tests should be marked and communicated to the customers.*

If additional information is needed for interpretation of the results, this should be added. The report may also contain information about deviations, statement of compliance, uncertainty, LOD, and opinions and interpretation. If the laboratory was responsible for the sampling, the test report should also contain all relevant information about this subject, such as the strategy and performance (see Standard section 5.7). Interpretation and opinions should be clearly marked as such in the report.

Electronic reports should also contain the parameters described above. Amendments to a test report should be made as a separate document that refers to the original report and should also meet the demands set.

If an additional or supplementary report is issued, this should be clearly identified as such.

17025:2005. The section numbers in these Boxes are the same as in the ISO/IEC Standard, but are presented in italics. The text that is taken from the standard is in normal font and is the 'simple' description of what is stated in the corresponding section in the standard. The text in italics is the additional related remarks. Both 'simple' description and related remarks are from the present authors, with the author's comments in italics.

4.4 A ROAD MAP FOR BUILDING A HIGH QUALITY SYSTEM

The previous sections describe the principles and elements of the QMS (in Section 4.2) and their linkages with the demands stated in Standard ISO/IEC 17025:2005 (Section 4.3 of this document). For accreditation, the laboratory must ensure all aspects are adequately implemented. The aim of this section is to describe a possible road map for the implementation of a QMS in four scenarios:

- A stand-alone feed analysis laboratory.
- A feed analysis laboratory in a large organization.
- A feed analysis research laboratory.
- A government or reference feed analysis laboratory.

Initially the accreditation of a stand-alone laboratory is described, and most of the steps required for such a laboratory are also valid for the other three types of laboratory. Subsequently, specific aspects and deviations from the first type of laboratory that need to be addressed for these three types of laboratories are then outlined.

Before starting, managers should realize that implementation is a continuous process that takes at least 3 to 4 years. During this period the system should be developed and improved. To make this process a success, a number of conditions need to be fulfilled:

- Upper management commits to the activity, with full cognizance of the importance of the QMS.
- Management provides resources required for implementation of the process.
- A realistic time schedule is prepared and implemented as per the plan.
- Prioritization of quality issues crucial for the results, and related implementation.
- Personnel are provided, especially technical staff, to support this process.

4.5 FIRST SITUATION: ROUTINE STAND-ALONE FEED ANALYSIS LABORATORY
4.5.1 Introduction

Standard ISO/IEC 17025:2005 is intended for routine laboratory work and is a perfect starting point for putting in place a QMS. Stand-alone laboratories operate in an open commercial situation and accreditation can be a powerful tool to gain customer confidence. This section describes a possible road map that leads to accreditation after a period of 4 years.

The first two years of this period should be spent on laying the foundations of a QMS based on ISO/IEC 17025:2005 for the laboratory. The focus of the first year should be on the technical aspects, while in the second year the focus is on organizational aspects. The last two years should be devoted to implementing all aspects of the Standard, with appropriate evaluation, with the aim to improve the QMS. A guideline outlining 'what to do when' is presented in Table 4.2.

Implementation of these aspects will be described in more detail in the following sections.

TABLE 4.2
Timetable for the implementation of the individual aspects of a QMS compliant with Standard ISO/IEC 17025:2005

Element	Start	Year 1	Year 2	Years 3 & 4
4.1		X[1]	X	
4.2			X	
4.3			X	
4.4		X		
4.5				X
4.6				X
4.7			X	
4.8		X		
4.9		X		
4.10			X	
4.11			X	
4.12			X	
4.13	X			
4.14			X	
4.15		X		
5.1	X			
5.2			X	
5.3		X		
5.4	X			X (5.4.6)
5.5		X		
5.6				X
5.7				X
5.8	X			
5.9	X			
5.10				X

Notes: "Elements" refer to the relevant sections in Standard ISO/IEC 17025:2005.
[1] Only Standard section *4.1.5(i)*.

4.5.2 Initial phase

Before the laboratory starts conducting analyses on a commercial basis the method description and validation (*5.4*), operational quality control (*5.9*), control of records (*4.13*), and handling of test items (*5.8*) should be available and operational. These are the minimum requirements to guarantee the quality of the results produced, and their implementation should be the primary priority for the laboratory.

At this time, the laboratory can base its methods, including sample preparation, on available documents, such as *Quality assurance for animal feed analysis laboratories* (FAO, 2011) or on internal specifications. Initially the validation can be limited to the determination of the LOD and LOQ, repeatability, and comparison with other laboratories. Operational quality control should be realized by obtaining standard sample(s) and the implementation of the first line of control (see Section 4.2.1.6 and Appendix B). All raw data that is essential for the calculation of results should be stored properly (see Appendix E).

4.5.3 First year

After becoming operational, the laboratory should further focus on technical aspects, such as facilities (*5.3*) and equipment (*5.5*). The laboratory should develop documentation regarding maintenance and performance checks (see Section 4.2.1.4 and Appendix F). Minimum requirements include the setting up of general procedures that guarantee the functionality of the laboratory as a whole; availability of specific equipment; and ensuring good housekeeping.

Other aspects the laboratory should implement are the reviews of requests (*4.4*), complaints (*4.8*) and control of non-conforming testing (*4.9*). These issues are related to its relationship with customers (commercial level) and will also help to increase their confidence in the laboratory.

The final aspect that should be implemented is the management review (*4.15*). It is important to involve upper management in this process right from the start, and to ensure that they are fully committed to the QMS. This review, after the first year, will be quite brief, and subsequently will become more comprehensive over time.

The laboratory should also initiate operational actions regarding the QMS during the first year. On the technical level, it should periodically evaluate its performance of tests and start participating in collaborative trials and proficiency tests. Experience shows that, especially in the initial period, the laboratory can benefit greatly from using this involvement and the resulting information to improve the quality of its tests.

The laboratory should also appoint a QA Manager who will be responsible for the implementation, evaluation, and improvement of the QMS (*4.1.5(i)*). For new laboratories, this person is often one of the laboratory technical staff with an affinity for quality issues, which is acceptable within the ISO/IEC 17025:2005 Standard. Experience, however, shows that quality issues will become more time consuming as time goes on, possibly leading to a fulltime position. This function also implies an independent organizational status, which means that the person will be placed outside the laboratory group (see Figure 3.4). During the first year, resources should be spent on training of the QA Manager, unless this person already has the requisite knowledge and expertise. The QA Manager may also have the role of Health and Safety Manager, or Environmental Manager, or both, or these might be the responsibility of another member of staff.

4.5.4 Second year

The main issues in the second year are the setting up and implementation of the QMS for the laboratory (organization level), and the evaluation and improvement of the quality of its technical performance.

The laboratory staff should start writing a quality control manual that describes the organization of the laboratory and its procedures. This manual is a general description of the total system and its documentation, and refers to more detailed underlying procedures. It should cover all aspects of the Standard (Standard section *4.2*). The Standard gives the laboratory freedom regarding the format of the manual. A simple approach is to format the quality control manual based on the different sections of the Standard, and describe how they are addressed. It can be advisable to hire external expertise to help with setting up and writing the quality control manual. Parallel to this writing, several decisions have to be made about the organizational structure of the laboratory, including all functions, responsibilities, authorization, and substitutes for vital management functions (Standard section *4.1*), registration of personnel and training records (Standard section *5.2*), and the distribution of documentation (Standard section *4.3*).

The writing of the quality control manual, including related procedures, is a process that is often complicated and time consuming, and demands good cooperation between the laboratory management and the QA Manager. It is important to emphasize that the laboratory management should decide how these aspects are organized and implemented, while the QA Manager should judge them against the demands of the Standard.

The QA Manager has an important responsibility in the process of document control to facilitate the setting up, labelling, release, distribution and review of documents. The laboratory should have a document control system that includes a distribution list for all current procedures, to ensure that only the most recent versions are available for personnel by removing all previous versions. This system can vary from simple spread sheets to a dedicated database. The standard offers the laboratory the opportunity to make small manual changes (hand-written amendments signed by authorized persons) to quality control documents, which can be a useful tool for the modifications frequently required in the initial phase. Review of documents and implementation of any manual changes should be undertaken regularly after consultation with the laboratory technical staff. Minor hand amendments should be made to all copies of controlled documents at the same time to avoid contradictory procedures. Documents should be reviewed and hand amendments made permanent as soon as is practicable.

During the second year, the technical performance of the laboratory should be evaluated. At a minimum, this evaluation should address the following parameters: results for the first line of control; repeatability; *z*-scores from collaborative trials; status of equipment; and complaints. The QA Manager should conduct this evaluation in the format of an internal audit (Standard section *4.14*). The QA Manager should set up a schedule of internal audits that describes items to be investigated and the frequency of audits. In this second year, the use of a checklist covering all aspects of the standard should be used, and deviations from the procedure should be recorded in audit reports (see Table 4.2, first data column). In response, an action report should be written by the laboratory that describes how the identified problems will be rectified, and includes a timeline to achieve this. The implementation of these actions should be overseen by the QA Manager.

The laboratory should describe and implement procedures for preventive and corrective actions (Standard sections *4.11* and *4.12*). These actions focus on avoiding the release of unreliable results. Corrective actions are consequences of non-conforming testing, deviations from internal audits, and complaints. The laboratory should follow an active policy of corrective actions to solve problems and shortcomings, and, if necessary, communicate problems to customers. Preventive actions aim to avoid future problems by monitoring trends in critical parameters that are indicators of the quality of the test. The laboratory should identify and regularly evaluate these parameters; some analytical knowledge is required for this. Examples of these parameters are any drift observed in control charts; values for blanks; and sensitivity analysis, such as the slope of calibration curves or peak areas. Trends can be graphed by using, for example, the CUSUM chart (see Appendix I) to provide a visual monitoring system. In an ideal situation this curve should vacillate around zero, whereas other shifts may be caused by systematic drift.

Preventive and corrective actions undertaken should be well documented, tested for efficiency, and signed. It is important that the laboratory can show that such procedures are clearly implemented in their routine operation.

The laboratory management should improve its service to customers by seeking their feedback (Standard section *4.7*). This can be achieved by sending an annual questionnaire to customers. The QA Manager should evaluate these results and use them as a tool to improve the laboratory's level of service and to identify opportunities for improvement.

The results from the internal audit and the responses from customers should be used in a systematic way to improve the QMS of the laboratory (Standard section *4.10*).

4.5.5 Third year

At the start of the third year, the laboratory should decide if it is ready to proceed with the official accreditation. If a positive decision is made, the laboratory should gain information about the most suitable accreditation organizations and contact them about the process. It is important that this organization is authorized for ISO/IEC 17025:2005 (i.e. be a signatory to ILAC, International Laboratory Accreditation Cooperation), which is not the case in every country. Most accreditation bodies will recommend an initial 'gap analysis' assessment followed by an initial assessment against ISO/IEC 17025:2005.

The third year should be devoted to implementing the remaining aspects that are more specific in the standard, and improving the existing QMS.

The remaining aspects on the technical level are sampling (Standard section *5.7*), estimation of the measurement of uncertainty (MU) (Standard section *5.4.6*), and traceability of measurements (Standard section *5.6*), for which procedures should be implemented. The set up and implementation of a procedure for sampling is only necessary if the laboratory conducts sampling for external clients, e.g. batch control of feed ingredients or diets, as part of their services. The laboratory should prove that it has sufficient analytical and statistical knowledge to conduct this in an appropriate manner.

Measurement of uncertainty should be estimated for all tests that will be accredited and although not required for non-accredited tests, it is good practice to also estimate for tests that are not covered by the accreditation. The laboratory should combine this estimation with an evaluation of the validation conducted in the first year. For accreditation it is necessary that all elements of the validation (see Appendix C) are estimated in a traceable way that is preferably based on an international standard. At this stage the laboratory should have enough knowledge and data to conduct these estimations. The results of these estimations can be used to calculate the measurement of uncertainty, which reflects the confidence interval (mostly 95%) of individual results for the different tests. Practice shows that laboratories can have serious problems with this aspect of the standard, which involves both analytical and statistical knowledge. International standards are available that can help laboratories to understand this requirement and perform the necessary calculations (see Appendix D). The calculated results should be realistic and available to customers.

For accreditation of a method, method validation and its relationship to an international standard method are important issues. Generally, the accreditation organization will accept the use of three types of methods, i.e. conforms to, based on or modification of, an in-house method. If a laboratory decides to use a method that 'conforms to' or is 'based on or modification of' a specific international standard method, it is obliged to prove that its performance matches the quality parameters of that international standard method. If the laboratory uses the 'in-house method', it does not have this obligation. However, it is imperative that all quality information, such as limit of quantification, precision and accuracy of the in-house methods, are made available to customers and users of the results. These demands, however, can vary between countries, and the laboratory should check this with the accrediting organization.

All these new values should be recorded in the analytical protocol for each test. These changes may be combined with a new standard format for these documents, although this is not a requirement of the standard.

Each piece of equipment requiring calibration should have calibration records that can be traced back to an International Reference, using reference standards, calibrated weights, thermometers, etc. For physical units, such as mass and temperature, calibrations should be conducted by accredited calibration providers to demonstrate traceability to reference standards. All relevant documents, including certificates, should be retained by the laboratory. Appendix H describes a traceable calibration for volumetric equipment that can be performed by the laboratory itself. Unless chemical calibration solutions are made under certified conditions, the laboratory has an obligation to prove its traceability by testing it against an independent solution (Standard section 5.6.3). This can be achieved by comparing solutions from different suppliers or against a solution prepared from pure chemicals under traceable conditions in the laboratory. The laboratory should explicitly describe the criteria for this comparison in its protocol.

The last few aspects that should be implemented are the procedures for subcontracting (Standard section 4.5), purchasing of services (Standard section 4.6), and reporting of results (Standard section 5.10).

For subcontracting, the laboratory should set up a clear selection process, based on objective criteria. Suitable laboratories for each test should be listed and made available to customers. The laboratory is obliged to evaluate the quality and service level of the laboratory used for subcontracting. It is important to emphasize that the laboratory remains responsible for the quality of the results in the case of outsourcing. These results should not be reported under accreditation.

A procedure and schedule should be available for the purchase of services and supplies, which also includes the evaluation of suppliers. For this purpose, the laboratory should make a list of critical chemicals and equipment and the specifications they should be tested against. The service level of the most frequently used suppliers should be regularly evaluated and documented. The QA Manager should take a leading role in this process.

Finally, the laboratory should ensure that the report of test results is confirmed against the demands stated in the standard. A possible approach is to implement a Laboratory Information Management System (LIMS) that automatically generates reports and meets standard specifications.

Internal audits should be conducted according to the previous mentioned schedule. Compared with the previous year, two changes will occur. The first change is the increase in

TABLE 4.3
Examples of two forms of audit report

Old form, very brief comments	New form, with more detail recorded
• Personnel records:	5.2.5 Personnel records:
The record of technician X was incomplete	The personnel records of technical staff X, Y and Z were examined. The records of Y and Z were complete. For technician X, the training and authorization file was not available.
• Calibration of equipment:	5.5.2 Calibration of equipment:
Calibration of balance X was not performed within specified time frame	The calibration of balances X, Y and Z, diluter A, and Kjeldahl equipment were examined. For all equipment, calibration procedures including specifications are available. The calibrations were performed in the previous year according to these procedures and judged against the specifications. Balance X was calibrated later than the stated deadline, without any action or registration: follow up on reason for delay.

the number of audits, which may require more people besides the QA Manager to undertake them. In this case, a select group of personnel should be trained to become internal auditors. It is important to realize that auditing contains a subjective element leading to the under- and overestimation of some issues depending on the importance an auditor gives to those issues. For this reason it is advisable to rotate auditors between the different laboratories or sections on a regular basis. The QA Manager is responsible for setting up a training programme to provide an in-depth knowledge of the standard for future auditors. The first audit should be conducted under supervision, after which they will be authorized if deemed competent by the QA Manager. Training records of all new auditors should be completed and stored in an appropriate manner. The QA Manager should initiate regular consultation between auditors to increase the quality and uniformity of the audit process. The standards for auditing management systems are in ISO 19011:2011.

The second change is the quality and detail expected in the audit reports. In the previous year, reports were mostly limited to checklists of investigated items and deviations found. The laboratory should now switch to a more detailed description of which items are investigated and what observations and deviations are found. The items should be described and labelled according to the Standard. Table 4.3 illustrates the differences between reports for two examples.

The management review written at the end of the year should be fully documented according to the demands of the standard (Standard section *4.15.1*). Compared with the previous years, this review should also contain information about its market positions (type and volume of work) and the preventive and corrective actions.

4.5.6 Fourth year

In the fourth year the laboratory should prove that all aspects of the QMS are implemented and operational according to the demands of the Standard. This year ends with an inspection from an official of the accreditation organization, and ultimately leads to the accreditation of the laboratory to ISO/IEC 17025:2005. To achieve this, the laboratory should monitor and improve the existing QMS.

The first step is to update the Quality Control manual and all procedures to ensure that all issues are sufficiently covered by these documents. For this purpose, the new Quality manual should also contain a reference table that links specific aspects of the standard to the appropriate sections in the Quality Control Manual. If in doubt, the laboratory can hire an external expert to check all its documents against the demands of the standard.

Internal audits to check the operational efficiency of the QMS should be conducted more frequently during this year. Observed deviations should be recorded and the relevant corrective actions completed. These audits should also pay attention to the presence of all relevant quality documentation, such as validation reports, personnel files, worksheets and calibration reports.

The laboratory manager and QA Manager should educate laboratory personnel regarding the quality procedures used and prepare them for future external audits. Experience shows that technical staff focus largely on the analytical work and have less inclination to follow the concept of a QMS in totality, which can lead to a difference between theory and practice. The QA Manager, supported by the laboratory management, should try to avoid this situation by explaining the QMS and the benefits of the different procedures. Performance of an audit by external expert(s) can be beneficial as preparation for technical staff, prior to the first official audit by the Accreditation Organization.

At the beginning of the fourth year, the laboratory should approach an authorized accreditation organization with the request for accreditation according to ISO/IEC 17025:2005. The laboratory should at this point make a definitive choice about which tests, including the matrix types, that are to be accredited. It is advisable to start with a limited number of tests and gradually increase them over the following years as 'extensions to scope'.

In general, the accreditation process will start with a pre-inspection of the laboratory with the aim to judge if it is capable of achieving the requirements of the standard, i.e. the presence of the facilities, equipment, Quality Control Manual, and personnel. After approval, an official inspection will be organized under the supervision of a team leader (Assessment Manager) who will focus on the general requirements (Chapter 4 of the Standard). Analytical activities (i.e. tests) are audited by technical experts, who will judge the quality of the results produced (Chapter 5 of the Standard). The number of technical experts will depend on the diversity of determinations that require accreditation. In most cases, these inspections will lead to a number of observed deviations (i.e. non-conformities) that should be addressed within a fixed period (usually 3 months). If all deviations are corrected in a satisfactory manner, the laboratory will be accredited for the tests specified.

The accreditation is always for a fixed term, generally four years. During this period, the laboratory will receive an annual maintenance inspection to judge different tests and aspects of the quality system. After four years, the whole system will be evaluated and judged by a new team of experts. After addressing all deviations and initiating corrective actions, accreditation will be granted for another period of four years.

After the accreditation, the laboratory should clearly inform customers which tests are accredited. This should also be marked in the final report of results (Standard section 5.10). Each year the laboratory can decide to add or remove tests from the accredited list or scope. New tests will be inspected during the next audit. It is crucial to inform the auditors, well in advance of the scheduled audit, if new methods are to be inspected. Methodology and validation reports will be required prior to this audit.

After achieving accreditation to ISO/IEC 17025:2005, the greatest challenge is to maintain and improve the QMS. A traditional pitfall is that the laboratory starts to relax on quality issues and focus on other things. The Accreditation Organization, however, expects the level of quality to increase with time, and this should be the prime objective of laboratory management.

4.6 SECOND SITUATION:
ROUTINE LABORATORY CONNECTED TO A FEED MANUFACTURER

The main difference between this type of laboratory and that described in the first situation is that it is part of a larger organization. For the implementation of a QMS based on ISO/IEC 17025:2005 it is important that the organization already has a quality system based on ISO standards (i.e. ISO 9001:2008). The presence of this system in the organization indicates the integrity of general procedures that are already implemented and operational, which the laboratory can use. An additional advantage is the presence of a quality control unit whose knowledge and experiences can be utilized to set up the QMS in the laboratory. These conditions give the laboratory a clear advantage compared with the first situation, and allows it to concentrate its attention more on the specific technical issues of ISO/IEC 17025:2005 (Chapter 5 of the Standard). Although there may be overlaps, it can be advisable to make a separate Quality Control Manual for the laboratory that refers to more general laboratory procedures. The advantage of this is that it concisely describes the total QMS of the laboratory, which makes specific inspections of the laboratory easier to perform. The presence of a general QMS can mean a reduction to three years for the period required to become accredited. Many of the requirements of an ISO 9001:2008 QMS will be shared with that of an ISO/IEC 17025:2005 QMS (or ISO 14001:2004 Environmental Management System or BS OHSAS 18001:2007 Occupational Health and Safety Management System) meaning that many SOPs may be shared, and form an Integrated Management System. It may be that if an organization has certification to ISO 9001:2008, ISO 14001:2004 or BS OHSAS 18001:2007, they might receive only one assessment visit from the certification body, who will assess against all three standards during the same assessment. Accreditation to ISO/IEC 17025:2005 will always be assessed separately by an accreditation body.

Integration of the laboratory within a larger organization increases its interaction with other sections and therefore increases the possibility of being influenced by others outside the laboratory. This issue is explicitly mentioned in Section *4.1.4* of the Standard as the potential for conflict of interests between the laboratory and production or commercial units. Upper management should guarantee and protect the independent position of the laboratory, which should be demonstrated in the structure of the laboratory and the authorization of its management. Upper management should demonstrate its commitment to the QMS by expressing this as a statement in the Quality Control Manual, by evaluation of management reviews, and being present at external audits.

4.7 THIRD SITUATION:
LABORATORY AS PART OF A RESEARCH ORGANIZATION

The main task of the laboratory in a research organization is to facilitate the research by conducting analytical activities. They are less focussed on routine analyses than in the first two situations. These laboratories have more flexibility in their scope of tests and there is a strong linkage with researchers, which may make an official accreditation harder to achieve. Some aspects of the standard, especially the technical issues such as avoiding improper influence (Standard section *4.1.4*) and the validation of in-house and non-standard methods (Standard sections *5.4.3*, *5.4.4* and *5.4.5*), can be useful to improve the quality of the analytical work.

Improper influence has already been discussed in the second situation and focuses on the independence and responsibility of the laboratory to guarantee an objective judgement of the results. In organizations where the laboratory is an integrated part of the research unit there is a strong interaction between technical staff and researchers. The greatest threat is that the researcher's expectations can influence the analytical judgement of technical staff. An example of this influence is illustrated in Table 4.4 which gives duplicate results for four samples. In the case of sample C, the difference was higher than permitted and the measurement was repeated leading to an additional value (i.e. 33.9). From an analytical point of view, the value 31.6 would be regarded as an outlier and the average of 33.9 and 34.7 should be the final results. The researcher however expects a linear relationship and will delete both values and use 31.6 as the final result.

Avoiding this type of interference is a challenge for research organizations where researchers try to find relationships to explain their observations. The pressure for commercial funding may increase this kind of interference, with the danger of such a subjective judgement of results.

A solution for this problem may be to limit direct interaction between technical staff and researchers and most importantly to ensure that the authorization of analytical methods and results will be conducted by the laboratory according to the set principles. The management of the unit should be aware of this problem and should implement mechanisms to avoid reporting of unreliable results. Examples of such mechanisms are the use of set procedures and criteria to evaluate results and conducting audits to evaluate conformity of the analytical work, both of which are described in the standard.

The second issue is to ensure prior validation of methods before they become operational. Laboratories within research organizations spend more time working on (new)

TABLE 4.4
Example of measurements

Sample	Measurements
A	10.0; 10.9
B	21.4; 20.9
C	31.6; 34.7; 33.9
D	40.5; 41.0

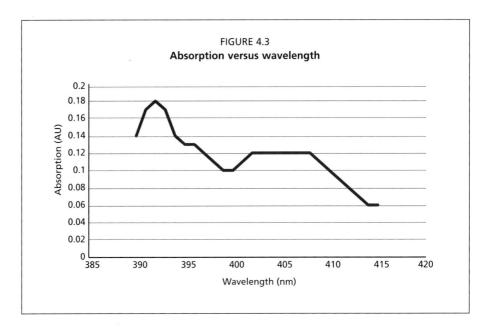

non-standard methods as an innovative aspect of the overall research work compared with the routine laboratory described in the first two situations. Introduction and implementation of these methods are also based on interaction with researchers and information available in the literature. Experience shows that these methods are often poorly validated and the expectations of researchers are used to evaluate the results, which can lead to unreliable results and conclusions. The laboratory can solve this problem by performing firstly an independent validation of the tests before use as stated in the standard (Standard section 5.4.4). The validation of these non-standard methods is however more complicated than for standard methods because the laboratory has an obligation to prove that the principle and the associated method are capable of measuring the analyte accurately. The laboratory should focus additionally on three issues: selectivity; sensitivity or interference in different matrices; and robustness (see Appendix A). In some cases, the laboratory should decide which quality parameters have the highest priority. For example, a higher sensitivity can negatively affect the robustness of the method, as illustrated in Figure 4.3, where the highest sensitivity is achieved at 392 nm, leading to the lowest values for LOD and LOQ. At this wavelength, however, the absorption is highly sensitive to small variations, whereas between 402 and 408 nm the absorption is more stable, although it is less sensitive, leading to a higher value for LOD and LOQ. In general, an improved robustness leads to lower variation in the results and improves the precision.

4.8 FOURTH SITUATION:
GOVERNMENT OR REFERENCE LABORATORIES

The important position of government or reference laboratories makes the implementation of a quality system based on ISO/IEC 17025:2005 mandatory. From the beginning, their analytical results should have the highest quality level to gain the full confidence of other laboratories. To achieve this level from the beginning is a great challenge because it does not give the laboratory time to build its own quality system that can be improved step by step as described in the first situation. In practice this means that all necessary validations and quality procedures should be fully implemented before the laboratory can start analysing samples for customers. To meet this challenge, the laboratory should attract quality manager(s) with suitable experience in this field. It may also be advisable to hire external experts to assist on special subjects, such as validation and calibration of equipment.

The laboratory should use the international standard methods and participate, if available, in international collaborative trials for each determination. This means that the laboratory should have state-of-the-art facilities and equipment, such as GC-MS, sequential in-line MS (MS-MS) and ICP-MS, for the determination of contaminants at trace level. Constant improvement of the quality of the analytical results and following international technical developments should be a vital part of their quality policy.

Sources used

ISO Standards

ISO 3534-2:2006. Statistics – Vocabulary and symbols – Part 2: Applied statistics. [Reviewed and confirmed 2010]

ISO 3696:1987. Water for analytical laboratory use – Specification and test methods. [Reviewed and confirmed 2011]

ISO 5725:1994. Accuracy (trueness and precision) of measurement methods and results.

ISO 6498. Animal feeding stuff – Preparation of test samples.

ISO 9001:2008. Quality management systems – Requirements.

ISO 10012:2003. Measurement management systems – Requirements for measurement processes and measuring equipment.

ISO 14001:2004. Environmental management systems – Requirements with guidance for use. [Reviewed and confirmed 2008]

ISO 19011:2011. Guidelines for auditing management systems.

ISO 21748. Guidance for the use of repeatability, reproducibility and trueness estimates in measurement uncertainty estimation.

ISO Guide 34:2009. General requirements for the competence of reference material producers.

ISO/IEC 17025:2005. General requirements for the competence of testing and calibration laboratories. [Reviewed and confirmed 2010]

Other sources

BSI. No date. BS OHSAS 18001:2007. Occupational Health and Safety Management System. See: http://www.ohsas-18001-occupational-health-and-safety.com/ohsas-18001-kit.htm Accessed 2013-08-03.

EURACHEM. 1998. EURACHEM Guide. The Fitness for Purpose of Analytical Methods. A Laboratory Guide to Method Validation and Related Topics. A document developed by a EURACHEM Working Group. See: http://www.eurachem.org/images/stories/Guides/pdf/valid.pdf Accessed 2013-08-03.

EURACHEM. 2007. EURACHEM/CITAC Guide: Use of uncertainty information in compliance assessment. First edition 2007. Edited by S.L.R. Ellison and A. Williams. See http://www.measurementuncertainty.org/pdf/Interpretation_with_expanded%20uncertainty_2007_v1.pdf Accessed 2013-08-03.

FAO. 2011. Quality assurance for animal feed analysis laboratories. *FAO Animal Production and Health Manual*, No. 14. Rome, Italy. Available at http://www.fao.org/docrep/014/i2441e/i2441e00.pdf Accessed 2013-08-29.

FAO. 2013. Quality assurance for microbiology in feed analysis laboratories. Prepared by R.A. Cowie and H.P.S. Makkar. *FAO Animal Production and Health Manual*, No. 16. Rome, Italy. Available at http://www.fao.org/docrep/018/i3287e/i3287e.pdf Accessed 2013-08-29.

ILAC-G17 2002. Introducing the Concept of Uncertainty of measurement in Testing in Association with the Application of the Standard ISO/IEC 17025. See: ILAC Web site www.ilac.org

JCGM 100:2008. [GUM 1995 with minor corrections]. Evaluation of measurement data – Guide to the expression of uncertainty in measurement / Évaluation des données de mesure – Guide pour l'expression de l'incertitude de mesure. Document produced by Working Group 1 of the Joint Committee for Guides in Metrology (JCGM/WG 1).

JCGM 200:2008. International vocabulary of metrology – Basic and general concepts and associated terms (VIM) / Vocabulaire international de métrologie – Concepts fondamentaux et généraux et termes associés (VIM). Available at http://www.bipm.org/utils/common/documents/jcgm/JCGM_200_2008.pdf. Accessed 2013-08-29.

OIML R 111-1 Edition 2004 (E). International Recommendation. Weights of classes E_1, E_2, F_1, F_2, M_1, M_{1-2}, M_2, M_{2-3} and M_3. Part 1: Metrological and technical requirements. International Organization of Legal Metrology. See: http://www.oiml.org/publications/R/R111-1-e04.pdf Accessed 2013-08-03.

Thompson, M., Ellison, S.L.R. & Wood, R. 2002. Harmonized guidelines for single-laboratory validation of methods of analysis (IUPAC Technical Report). *Pure and Applied Chemistry,* 74(5): 835–855.

VIM *See:* JCGM 200:2008.

Appendix A
Ensuring quality analytical performance

A1. DETERMINATION OF LIMIT OF DETECTION (LOD) AND LIMIT OF QUANTIFICATION (LOQ)

LOD is the lowest analyte concentration that can be reliably distinguished from the blank, while LOQ is the lowest concentration at which the analyte can be reliably detected and also at which predefined goals for accuracy and precision can be met.

The internationally accepted formulas for the estimation of the LOD and LOQ are based on the signal of a blank sample (i.e. without the analyte) and its standard deviation (SD)[1]

$$LOD = blank + 3 \times SD \text{ (blank)}$$
$$LOQ = blank + 10 \times SD \text{ (blank)}$$

Examples of LOD and LOQ calculations for different method types are given in the following sections.

A1.1 Gravimetric analyses

The LOD and LOQ of gravimetric determinations depend on the accuracy of the balance used, which is often specified by the supplier of the balance. The accuracy of a good quality four-figure balance should be approximately 0.3 mg (supplier specification).

This accuracy is equal to the 95% confidence interval, which is the average ±2 × SD (standard deviation). The range of this confidence interval or the accuracy equals 4 × SD, which means if the accuracy is 0.3 mg (in this case) the SD is 0.075 mg (i.e. 0.3/4).

For balances, the signal of a blank can be set to zero.

Using the general formulas:

$$LOD = 0 + 3 \times 0.075 = 0.225 \text{ mg}$$
$$LOQ = 0 + 10 \times 0.075 = 0.750 \text{ mg}$$

The LOD and LOQ for determinations are generally expressed on a content basis and therefore depend on the amount of sample used. For example, if 1 g of sample is used, using the above example, the LOD will be 0.225 g/kg (i.e. 0.225 mg/1 g) and the LOQ will be 0.750 g/kg (i.e. 0.750 mg/1 g).

The LOD and LOQ can be improved, i.e. values for LOD and LOQ can be lowered, if required for more sensitive measurements by using more sample material or a more accurate balance. The above-stated LOD and LOQ values for the gravimetric determinations, such as dry matter, ash, fat and fibre, are generally sufficient to analyse feedstuffs without the need for further improvement.

---

[1] **MacDougall, D & Crummett, W.B.** 1980. Guidelines for data acquisition and data quality evaluation in environmental chemistry. *Analytical Chemistry*, 52(14): 2242–2249.

If the accuracy information is not available from the supplier of the balance, the laboratory should estimate the standard deviation of the balance by measuring a weight several times (at least ten times) over several days.

A1.2 Titration-based analyses

The LOD and LOQ of titration-based analyses, such as the Kjeldahl determination for nitrogen (N), depend on the value found for the blank and the standard deviation of the blank. For Kjeldahl determinations, the blank is the digestion solution, including catalysts but omitting the sample. These blanks should be analysed at least six times, and used as in the following calculation.

An example:

Values observed (ml) are 0.15; 0.18; 0.13; 0.14; 0.18; 0.17, which leads to an average of 0.158 ml and a standard deviation of 0.021 ml.

Using the general formulas:

$$LOD = 0.158 + 3 \times 0.021 = 0.222 \text{ ml}$$
$$LOQ = 0.158 + 10 \times 0.021 = 0.372 \text{ ml}$$

To express the LOD and LOQ as content in the sample, the following formula for the determination of N should be used:

$$N(g/kg) = (V - V_{blank}) \times M_{(acid)} \times f \times 14/W$$

Where: V is the volume of the acid used (i.e. 0.222 ml for LOD and 0.372 ml for LOQ)
V_{blank} is the volume of the acid used by the blank (i.e. 0.158 ml)
M(acid) is the molarity of the acid used (i.e. 0.1 M HCl)
f is the acid factor (i.e. 1 for HCl)
14 is atomic weight of Nitrogen (N)
W the weight of the sample (i.e. 1 g)

For LOD: N = (0.222 - 0.158) × 0.1 × 1 × 14/1 = 0.09 g/kg
For LOQ: N = (0.372 - 0.158) × 0.1 × 1 × 14/1 = 0.30 g/kg

A1.3 Spectrometric analyses

The LOD and LOQ depend on the absorption value found for the blank and the standard deviation of the blank. The blank is the diluted reagents or colour reagents, without the analyte. These blanks should be analysed at least six times and used as in the following calculation.

An example:

Observed absorption units (A.U.) are 0.004; 0.005; 0.003; 0.003; 0.006; 0.008, which leads to an average of 0.004 A.U. and a standard deviation of 0.002 A.U.

Using the general formulas:

$$LOD = 0.004 + 3 \times 0.002 = 0.010 \text{ A.U.}$$
$$LOQ = 0.004 + 10 \times 0.002 = 0.024 \text{ A.U.}$$

To express the LOD and LOQ as the concentration of the analyte in the measured solution, the calibration curve should be used. For example, the calibration curve is:

$$\text{Absorption (A.U.)} = 0.500 \times \text{concentration (mg/L)}$$

The LOD in the measured solutions is 0.020 mg/L (i.e. 0.010/0.500) and the LOQ is 0.048 mg/L (i.e. 0.024/0.500).

To express the LOD and LOQ as a content of the sample, the following formula should be used, and in general for spectrophotometric determinations, the formula can be expressed as:

Content (g/kg) = concentration (mg/L) × volume (L) × dilution factor/sample weight (g).

Where:
>Concentration is the content of the analyte in the measured solution.
>Volume is the end volume of the extracted solution
>Dilution factor is the additional dilution needed to get the value of the measured solution within the range of the calibration line.
>Weight is the amount of sample used.

For example, if 1 g of sample is used, the end volume is 0.100 L, and the dilution factor is 25, then

$$LOD = 0.020 \times 0.1 \times 25/1 = 0.05 \text{ g/kg}$$
$$LOQ = 0.048 \times 0.1 \times 25/1 = 0.12 \text{ g/kg}$$

LOD and LOQ for spectrometric methods can be improved by increasing the sample amount and decreasing the volume and dilution factor.

A1.4 CHROMATOGRAPHIC ANALYSES

Different methods for the determination of the LOD and LOQ for chromatographic methods are available. A good approach is to analyse an analyte at a near background level and to determine the standard deviation at the peak area. This value should translate to concentrations and content by using the same approach as for the spectrometric method.

An example:
Values found (Area) are 2500; 2900; 3000; 2700; 2400; 2700, which give an average of 2700 area and a standard deviation of 228 area. Using the general formulas:

$$LOD = 3 \times 228 = 684 \text{ Area}$$
$$LOQ = 10 \times 228 = 2280 \text{ Area}$$

To express the LOD and LOQ as a concentration of the analyte in the measured solution, the calibration curve should be used. For example, if the calibration curve is:

$$\text{Area} = 100\,000 \times \text{concentration (mg/L)}$$

the LOD in the measured solution is 0.00684 mg/L (i.e. 684/100 000) and the LOQ in the measured solution is 0.0228 mg/L (i.e. 2280/100 000).

To express the LOD and LOQ as a content of the sample, the formula used to calculate content should be used:

Content (g/kg) = concentration (mg/L) × volume (L) × dilution factor/sample weight (g)

Where:
>Concentration is the content of the analyte in the measured solution.
>Volume is the end volume of the extracted solution.
>Dilution factor is the additional dilution needed to get the value of the measured solution within the range of the calibration line.
>Weight is the amount of sample used.

For example, if 0.5 g sample is used, the end volume of the measured solution is 0.050 L, with a dilution factor of 10:

$$LOD = 0.0068 \times 0.050 \times 10/0.5 = 0.0068 \text{ g/kg} = 6.8 \text{ mg/kg}$$
$$LOQ = 0.022 \times 0.050 \times 10/0.5 = 0.022 \text{ g/kg} = 22 \text{ mg/kg}$$

To improve the LOD and LOQ, the same approach can be used as for spectrometric determinations. Chromatographic methods also offer additional opportunity through increasing the injection volume used, which will increase the sensitivity or slope of the calibration curve.

A2. LINEARITY

Linearity is generally determined by using a calibration curve of the analyte and extrapolating to predict the value of the next calibration point. This value should be compared with the measured value and the difference should not be larger than a fixed value (generally 5%).

An example:

For calibration solutions of 2, 4, 6, 8 and 10 mg/L the measured absorption values are 0.203, 0.405, 0.608, 0.790 and 0.930. The first step is to calculate the calibration line based on linear regression for the first three points (i.e. 2, 4 and 6 mg/L), which leads to the curve:

$$\text{Absorption} = 0.1012 \times \text{concentration (mg/L)} + 0.0003.$$

The next step is to predict the absorption value for 8 mg/L, which is 0.810 (i.e. $0.1012 \times 8 + 0.0003$). The difference from the measured value is 0.020 (i.e. 0.810 −0.790) or 2.5%, which is within the acceptable range (<5%). Therefore the linearity can be extended to 8 mg/L and the calibration curve is calculated for the first four points, which leads to the curve:

$$\text{Absorption} = 0.0982 \times \text{concentration (mg/L)} + 0.010.$$

The next step is to predict the absorption value for 10 mg/L, which is 0.992 (i.e. $0.0982 \times 10 + 0.010$). The difference from the measured value is 0.062 (i.e. 0.992 −0.930), or 6.2%, which is outside the acceptable range (>5%). Therefore the linearity is limited to 8 mg/L for the measured solution.

For results above this concentration, the measured solution should be diluted so that the concentration is <8 mg/L.

A3. PRECISION

The precision of a determination is expressed as the repeatability and the intra-laboratory reproducibility, which are related to the standard deviation (SD). The repeatability equals by definition $2\sqrt{2}$ (i.e. 2.8) × SD if the sample is measured in the same batch (equal conditions). The intra-laboratory reproducibility equals by definition $2\sqrt{2}$ (i.e. 2.8) × SD if the sample is measured in different batches (non-equal conditions). Table A1 below shows a typical example of results obtained for the determination of the precision. The same sample is measured in triplicate on three different days. For each batch, the average and SD is estimated and used to calculate the relative SD (i.e. standard deviation × 100%/average) and the relative repeatability (i.e. relative SD × $2\sqrt{2}$). The relative repeatability of the batches varied from 2.92 to 3.42%, leading to an average value of 3.4%.

The intra-reproducibility is calculated using all values from the different batches (in this case nine values) leading to an average of 39.8 and an SD of 0.79 (see last line of Table A1). The relative SD is 1.98, leading to a relative intra-laboratory reproducibility of 5.55% (i.e. $2\sqrt{2} \times 1.98$) and is in this case 5.5%.

TABLE A1
Calculation of precision

	Measured values			Average	Standard deviation (SD)	Relative SD	Relative repeatability
Batch 1	40.3	40.1	39.5	39.97	0.42	1.04	2.92
Batch 2	38.3	39.2	39.3	38.93	0.55	1.41	3.96
Batch 3	41.0	40.2	40.1	40.43	0.49	1.22	3.42
Total				39.8	0.79	1.98	5.55

The value for repeatability is normally lower than for the intra-laboratory reproducibility because it involves fewer sources of variation.

Generally, laboratories analyse duplicates in the same batch and should use the relative repeatability as part of their acceptance criteria. The intra-laboratory reproducibility could be used as criteria for samples, such as the control sample, and this should be analysed in every batch.

A4. ACCURACY

Accuracy can be tested by analysing certified reference materials (CRM) or by collaborative trials (performance testing). In both cases, the z-score could be used to evaluate the result. The z-score is the difference between the measured and the stated value expressed in units of the SD.

For CRMs, the stated value is the reference value given by the supplier, which is mostly based on a collaborative trial using various techniques. The laboratory should compare this value with the measured value and divide the difference by its intra-laboratory reproducibility for that method.

An example:
Measured value is 10.0 mM and the relative SD is 2%, or 0.2 mM in this case. The reference value is 9.5 mM. The calculated z-score is:

$$Z\text{-score} = |\,10.0 - 9.5\,| / 0.2 = 2.5$$

For collaborative trials, the laboratory should compare its result with the average and divide the difference by the standard deviation of the collaborative trial. In most cases, the organizer will automatically do this calculation and report the z-scores for each individual participant.

A z-score <3 means that the measured value is within the 99% confidence interval (i.e. ±3×SD) or the reference value (in case of CRM) or the average of the collaborative trial. In this case, no systematic error is observed, meaning that the accuracy of the test is sufficient. If the z-score is >3, there is a systematic error and the laboratory should try to identify the cause and rectify the problem. Different methods, such as modifying the protocol or the use of an internal standard, are available to improve the accuracy. Information about the accuracy should be used in the calculation of the uncertainty of measurement (see Appendix D), and mentioned in any communication with customers.

A5. SELECTIVITY AND SENSITIVITY

Selectivity means that the measured value is only caused by the specific analyte. Consequently, the absence of this analyte should lead to no detectable signal. However, this is often not the case. Gravimetric methods are non-selective and mostly based on assumptions, as in the case of crude ash, fat and fibre. For chromatography, the selectivity is based on the quality of the separation and the type of detector, which means that only detectable analytes can be investigated. The use of a more selective detector, such as fluorescence, limits the list of compounds that can interfere compared with UV detection. A list of possible interfering compounds should be analysed with the method and evaluated.

Sensitivity is the response of the measured value relative to the concentration of the analyte, and is mostly expressed as the slope of a calibration line, or the response factor in case of a one-point calibration. Other components can influence the sensitivity in two ways. Firstly, their presence can lead to a background signal, which is often found for colour-based spectrometric determinations. The results should be corrected for this background if this value is always the same. If these values differ, individual corrections should be made by measuring the differences (as for kinetic methods). An example of this approach is:

Calibration solution:	Start: 0.004	End: 0.250	$\Delta A = 0.246$
Sample solution:	Start: 0.069	End: 0.149	$\Delta A = 0.080$

Both ΔA values are used for the calculation of the analyte in the sample.

Secondly, the matrix of the sample can interfere with the determination of the analyte and therefore affects the sensitivity. This effect is demonstrated in Table A2, showing the responses to the same analyte at different levels, in different matrices.

The results show that the response in water is high when compared with acid. This means that if the samples have an acid matrix, the use of a calibration curve in water will lead to an underestimation of the content of the analyte. This problem can be observed in the determination of minerals by AAS or ICP and colour-based spectrometric determinations, and can be solved by matrix matching, i.e. the matrix of the samples and those of the calibration solutions should be the same.

A particular problem occurs when there is a matrix difference between samples, leading to different response factors for the analyte. In this case, the method of standard addition

TABLE A2
Response variation with concentration of analyte

Analyte (ppm)	Absorption (A.U.) in	
	Water	Acid
0	0.000	0.000
1	0.100	0.090
2	0.200	0.180
3	0.300	0.270
Slope	0.100	0.090

Appendix A – Ensuring quality analytical performance

to each sample should be applied. The method of standard addition is based on adding a fixed amount of the analyte to the sample and using the increase in the response to calculate its original content.

An example:
A liquid sample is presented with an unknown content a, of analyte X.
First the sample will be diluted in two ways.
- Dilution #1: 1 ml sample + 1 ml solution without analyte X. The content of analyte X in this solution becomes $0.5 \times a$. The measured absorption of this solution was 0.200 A.U.
- Dilution #2: 1 ml sample + 1 ml solution containing 1 mg analyte X (content is 1 mg/ml). The content of analyte X in this solution becomes $0.5 \times a + 0.5$ (i.e. the content of the diluted sample plus the content of the diluted solution containing analyte X). The measured absorption of this solution was 0.300 A.U.

The difference in absorption between both solutions is 0.100 A.U., which is due to the difference in concentration of 0.5 mg/ml (i.e. $(0.5 \times a + 0.5) - 0.5 \times a$). Therefore a concentration of 0.5 mg/ml leads to an absorption of 0.100 A.U. The absorption of dilution #1 was 0.200 A.U. which corresponds to a concentration of 1.0 mg/ml, which equals $0.5 \times a$. Therefore a, or the content in the original sample, is 2.0 mg/ml.

A6. ROBUSTNESS

Robustness is the effect of variation in the conditions on the final result, such as differences in temperature and time during incubations, extractions, drying and incineration steps. These effects can be investigated by performing the test and varying these conditions, such as drying at 105°C instead of 103°C. The differences can be expressed as a z-score to evaluate if they are significant.

An example:
Dry matter of a control sample is 910 g/kg with an SD of 1 g/kg when drying at 103 °C. Table 3 shows the results for the control sample obtained at different temperatures.

The z-scores at 102 ° and 104 °C are 2, which means that these results are within the 99% confidence interval of the value found at 103 °C and do not lead to a significant difference in the result. The z-scores at 101 and 105° C are 4, which means that these results are outside of the 99% confidence interval of the value found at 103 °C, and therefore significant. In this case the drying temperature should be between 102 and 104 °C, which should be described in the protocol as 103 ±1 °C.

TABLE A3
Effect of temperature variations on dry matter results

Temperature (°C)	Dry matter	z-score
101	914	4
102	912	2
104	908	2
105	906	4

Appendix B
First line of quality control and the use of Shewhart charts

B1. FIRST STEP: CHOICE OF THE CONTROL STANDARD

The principle of this procedure is based on the assumption that the deviation in the results found whilst using the control standard is similar to that of individual samples. To fulfil this assumption, the matrix and content of the control standard(s) should be comparable with the samples to be analysed. This comparison is also necessary because the results of the control standards are used to calculate the precision and uncertainty of measurement for samples.

For animal nutrition laboratories, the most obvious choices for control samples are feed ingredients or complete diets. Special attention should be focused on the level of individual nutrients in these products. Maize for instance, cannot be used as a control standard for fat, whereas fat-rich soybeans can be used for starch.

The physical requirements of the control standard are stability and homogeneity, which allows results to be compared over time and between batches (see third step, below). The requirements of a control sample are primarily precision and repeatability, rather than accuracy. The laboratory can produce its own control sample at relatively low cost.

B2. SECOND STEP: PREPARATION OF THE CONTROL STANDARD

The laboratory can produce its own control standard as long as the requirements regarding stability and homogeneity are fulfilled. The standard procedure is to select a large batch (at least a few kilograms) of material (such as a feed ingredient or diet), dry, and grind to a specific size (usually 1 mm). The material should be divided into smaller portions of approximately 300 g by using a splitter or a rotation sample divider (see ISO 6498). These portions should be numbered and stored in plastic bags or bottles under appropriate conditions (generally at +2 to +8 °C).

Particularly in the case of contaminants, the laboratory can create control standards by addition of a fixed amount of the analyte(s) to be determined into the control standard material. Special storage conditions (low temperature, under vacuum, in the dark, etc.) may be required if the analyte to be measured is unstable.

B3. THIRD STEP: DETERMINATION AND CONTROL OF HOMOGENEITY OF THE CONTROL STANDARD.

The level of each analyte for which the control standard will be used should be determined by chemical analysis. This determination can be combined with the test for homogeneity of the control standard. For this purpose, the square root of the number of bags or bottles (in which the sample was distributed) should be analysed in triplicate at least. These initial samples should be divided at the time of the sampling process. Results should be statistically analysed by Analysis of Variance (ANOVA) to show that the variations within portions are not significantly different. After passing this test, the results should be combined to calculate the overall average and SD, and these form the basis for the limits in the first control or Shewhart chart.

B4. FOURTH STEP: SET UP AND USE OF THE SHEWHART CHART

The Shewhart chart is designed based on the results obtained from the previous step and contains five lines:
- The first line represents the average value found for the analyte.
- The second and third lines are the boundaries of the 95% confidence interval. The values for these boundaries are calculated by average $-2 \times$ SD (lower line) and average $+2 \times$ SD (upper line).
- The fourth and fifth lines are the boundaries of the 99% confidence interval. The values for these boundaries are calculated by average $-3 \times$ SD (lower line) and average $+3 \times$ SD (upper line).

An example:
If the average is 100 g/kg and the standard deviation is 5 g/kg, the first line (average value) will be at 100 g/kg, the second and third 95% confidence lines will be at 90 g/kg (i.e. 100 -2×5) and 110 g/kg (i.e. 100 $+2 \times 5$), and the fourth and fifth (99% confidence) lines will be at 85 g/kg (i.e. 100 -3×5) and 115 g/kg (i.e. 100 $+3 \times 5$). This means that if considering only random variation, 95% of all results should be between 90 and 110 g/kg and 99.5% of all results should be between 85 and 115 g/kg.

In this chart, the results of the control standard found in the analysed batches are chronologically drawn, as illustrated in Figure B1. The middle line represents the average, the lines above and below that are the boundaries of the 95% confidence interval, and the uppermost and lowermost lines are the boundaries of the 99.5% confidence interval. The dots represent the results found for the control standard in the batches.

The quality of the results found in different batches are judged based on the result of the control standard in that batch. Although there are no strict internationally accepted criteria for this judgement, the following criteria are generally used to identify batches that do not meet the acceptance specifications:
1. Outside the 99.5% confidence interval, which means in this case a content lower than 85 g/kg or higher than 115 g/kg.
2. Twice in sequence outside the same side of the 95% confidence interval, which means in this case twice between 85 and 90 g/kg or twice between 110 and 115 g/kg.
3. More than ten times in sequence on the same side of the average.

These criteria are selected based on the chance that this situation occurs in the case of random variation. In the first situation, this chance is 0.5%, or 1 in 200 batches, whereas in the second and third situations these chances are even smaller. These situations are therefore a strong indication of a problem within the analyses of that batch, which could lead to unreliable results. In Figure B1, this situation occurs in batch 2 (outside the 99.5% confidence interval), batch 6 (twice outside on the same side of the 95% confidence interval), and batch 17 (more than 10 times on the same side of the average). The results obtained

Appendix B – First line of quality control and the use of Shewhart charts

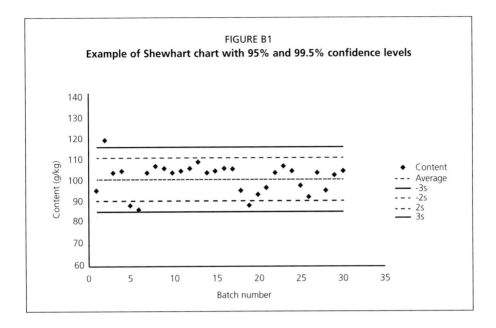

in these batches are unreliable and should be rejected, and action should be taken. In most cases, the batch will initially be repeated. If the results for the control standard remain outside the expected control values, the test should be stopped and a more structured investigation instigated (see Standard sections *4.9* and *4.10*).

After 30 measurements, or after a specific time interval, the average, the standard deviation, and consequently the boundaries of the confidence intervals, should be re-calculated based on previous and new data. A chart with new lines should be drawn and used for the next values for the control standard. In more sophisticated approaches, an F-test and a t-test can be performed to check the equality between the old and new Stewart's chart.

Appendix C
Validation requirements

For all tests, the laboratory should first set up and conduct a validation study.

The first step is to define the matrices and typical concentration level of the analyte, which should be based on the routine samples the laboratory usually analyses. Although a smaller number of matrices will reduce the time spent on the validation, the laboratory should note that this also reduces the types of sample that it can analyse with this test.

The second step is the choice of the method. For validation requirements, a distinction should be made between standard methods that are officially published, and laboratory-developed (i.e. in-house) or non-standard methods.

C1. STANDARD METHODS
The validation study should contain a determination of the following items:
- Limit of Detection;
- Limit of Quantification;
- Accuracy; and
- Precision (repeatability and reproducibility).

If the method uses a calibration standard curve, the linearity should also be examined.

The values found can be used in two different ways. Firstly, these values can be compared with those mentioned in the standard method. Values must be equal to the standard values as a minimum requirement. The laboratory can also claim conformity if its performance is better than that stated by the standard method.

Secondly, the laboratory may use values from the validation study to describe the quality of the test without comparing them with those from the standard method.

C2. NON-STANDARD METHODS
Aside from the items mentioned in the standard methods, a validation study should also address selectivity, sensitivity and robustness. The laboratory should use the results from the validation study to describe the quality of the test.

Appendix D
Calculation of uncertainty of measurement

The uncertainty of measurement is the dispersion of the value that is related to the concentration of the substance being measured. This dispersion is mostly expressed as a 95% confidence level around the value found. The principle of the calculation of the uncertainty is described in *Eurachem/CITAC guide. Quantifying uncertainty in analytical measurement* (available at: www.eurachem.org/index.php/publications/guides/quam), which also gives examples to illustrate this complex measurement. The aim of this present appendix is to give some general information that can help laboratories to understand the principle and perform the calculations.

Methods can be divided into two groups:
1. **Rational methods:** the values of the analyte should be independent of the method used. Examples are mineral analyses or individual organics, such as amino acids and fatty acids.
2. **Empirical methods:** the value of the analyte depends on the method used. Examples are the determination of crude protein, crude fat and crude fibre.

Rational methods can contain a bias or systematic error leading to a difference for the true value, whereas for empirical methods this bias is by definition absent.

There are two approaches to calculate the total uncertainty:
1. Evaluate the uncertainty from each individual source and combine them. This is a theoretical approach.
2. Estimate the combined uncertainty from method performance data by combining only those individual sources of error if they contribute to more than one-third of the total uncertainty.

The second approach is closer to actual practice and is therefore preferable. This method however requires performance data, and a certain time period before it can be performed accurately. This approach will now be discussed in more detail. In general the uncertainty is calculated from:
- Estimation of the precision.
- Bias study, which can be performed by analysing certified reference material (CRM), comparison with a reference method, or by measuring the percentage recovery.
- Additional factors outside the standard analyses that can contribute to the uncertainty, such as sampling or storage.

For empirical methods, the bias is considered to be zero and the laboratory should control the method parameters, such as time or temperature. For rational methods, the laboratory should estimate the significance of the bias compared with the combined uncertainty. If this value is significant, the laboratory should make a correction or report this value to

TABLE D1
Example of the estimation of the uncertainty

Source	Average	Uncertainty	Relative uncertainty
Precision	50	5	0.1
Bias	45	4.5	0.1
Combined			0.14
Expanded			0.28

the customer. Information obtained from collaborative trials or CRM, can for both method types be used directly for the estimation of the uncertainty.

The combined uncertainty is calculated as the root of the sum of the squares of the individual uncertainties. This combined uncertainty is the SD of the determination, and covers all possible sources of variation in its result. The 95% confidence interval is calculated by multiplying the combined uncertainty by a factor of 2, which gives the expanded uncertainty and is commonly reported. The uncertainty is expressed as a maximum of 2 digits, which also determines the number of digits for the result.

The contribution for precision can be estimated by performing duplicate analyses and calculating the SD of the normalized differences (difference divided by the mean) and divide by $\sqrt{2}$. The contribution for bias can be estimated by performing recovery experiments and calculating the average SD, which should be divided by $\sqrt{(\text{number of measurements})}$.

The calculation of the combined and expanded uncertainty is given in Table D1.

In this case, the relative expanded uncertainty is 0.28, which means that the 95% confidence interval (2 × 0.14) of the measured value is the value ±0.28 × value. Suppose that the result is 200 g/kg, then the 95% confidence interval is between 144 (i.e. 200 −0.28 × 200) and 256 g/kg (i.e. 200 +0.28 × 200).

Additional individual sources should be evaluated if they contribute more than one-third of the combined uncertainty, which in this case is approximately 0.05.

Appendix E
An example of technical records for a determination

Technical records should contain all relevant information that is necessary for traceability of the performance of the test and the end results. A standard format for the determination of nitrogen in samples is given below.

Determination of N according to   code of the test
List number: NNN
Date: DD/MM/YYYY
Technician/Operator: name
Titrant: 0.05 M H_2SO_4
Volume blank (V_0):   0.15 ml

Sample list

No.	Label	Weight (g)	V1 (ml)	V1 -V0 (ml)	N (g/kg)1
1	2012-100	0.7012	5.34	5.19	10.4
2	2012-101	0.6934	6.72	6.47	13.1
3	2012-102	0.7056	14.99	14.84	29.4
4	2012-103	0.6912	18.74	18.59	37.7
5	2012-104	0.7134	15.43	15.28	30.0
6	2012-105	0.6911	8.79	8.64	17.5
7	2012-106	0.7123	12.34	12.19	24.0
8	2012-107	0.6975	4.35	4.20	8.4
9	2012-108	0.7044	9.78	9.63	19.1
10	Control standard	0.6943	13.45	13.30	26.8

Notes: (1) N (g/kg) calculated as ($V_1 - V_0$) × titrant (0.05) × 2 × 14/weight

Technician/Operator: *Signature of technician/operator*
Control by: Name of authorized person
Approved: Yes or No
Remarks: Special observations or other important issues
Date: *Date of control*
Signed: *Signature of authorized person*

Appendix F
An example of a maintenance and calibration document

A possible format for a maintenance and calibration document could be:

MAINTENANCE AND CALIBRATION DOCUMENT

Category:	Analytical balance
Manufacturer:	Name of manufacturer
Type:	Model of balance
Serial number:	Identification Number from manufacturer
Identification code:	Registration code given by the laboratory
Position:	Registration code of the laboratory room

USER MAINTENANCE AND CALIBRATION

Daily performance
- Clean outside of equipment.
- Perform internal calibration check.

Every three months
- Calibration using mass weights of *NNN* gram. Measured values should be within ±*N.N* mg of the reference value.
- Supplier maintenance and calibration.

Annually
- Cleaning and check of equipment, including electrical safety.
- Calibration of balance traceable to International Standards. Measured values should be within ±*N.N* mg of the reference value.
- Adjustment of equipment in case of deviation.

Appendix G
An example of a training record

TRAINING RECORD

Name of staff member: *Name*
Name of supervisor: *Name*
Determination: *Description of the test (internal code)*
Training activities:
Explanation and discussion about the method, including the use of the equipment and control standards
Analysis of list XX under supervision:
Differences between duplicate analyses were within the specifications. Results for the control standard were within the specification (see appended sheet for data).
Performance without supervision of a list of control standards. Nine out of 10 within specifications (see appended sheet).

       Authorization of   *Name of the staff member* from *date.*

       Signed by   *Signature of supervisor*

       Date: DD/MM/YYYY

Notes:
1. Different categories of training status can be included, e.g. "has read and understood the method", "some experience with supervision", "capable of performing the method without supervision" or "ability to train others".
2. Attach appended sheets.

Appendix H
Procedure for traceability of volumetric calibration

This procedure describes a method to perform a traceability calibration of volumetric equipment used by the laboratory. In practice, this could be used for diluters, pipettes and volumetric flasks.

General principle
- Boil demineralized water to remove CO_2 gas and allow it to cool down to room temperature.
- Determine the absolute temperature of the water with a calibrated thermometer (take into account the measured uncertainty in the thermometer).
- Determine the mass of the volume of water obtained by the tested volumetric equipment. For this purpose, use a certified balance (take into account the measured uncertainty in the balance).
- Evaluate the results against the described specifications.

An example
Calibration of a diluter with fixed syringes of 1.0 ml and 5.0 ml.
  Specifications: Deviation of volume of syringes <0.5%.
  Temperature of the boiled water, after cooling: 20.0°C
  Systematic deviation of thermometer used against reference: +0.5°C
  Temperature of boiled water after correction: 19.5°C
  Density of water at 19.5°C: 0.9983 g/ml
For 1.0 ml syringe:
  Disperse 5 times the volume of the syringe and record weight.
  Weight: 4.9958 g
  Systematic deviation of balance used against reference: -0.0003 g
  Corrected mass: 4.9961 g
  The volume should be calculated by the mass divided by the density, which equals 5.0046 ml (i.e. 4.9961 g/0.9983 g/ml). The deviation against the stated value (i.e. 5.0000 ml) is 0.09%, which is within specifications.
For 5.0 ml syringe:
  Disperse 5 times the volume of the syringe and record weight.
  Weight: 24.7861 g
  Systematic deviation of balance used against reference: -0.0003 g
  Corrected mass: 24.7864 g
  The volume should be calculated as the mass divided by the density, which equals 24.8286 ml (i.e. 24.7864 g/0.9983 g/ml). The deviation against the stated value (i.e. 25.0000 ml) is 0.68%, which is outside specifications.
  In this case, further action is needed.

Appendix I
Trend analysis

Cumulative sum (CUSUM) charts are a powerful tool to detect small shifts in the mean of a process and are more suitable for this purpose than the Shewhart charts described in Appendix B. In these charts the deviation for the approved average is cumulative, plotted against time and therefore visible if a drift appears in the assay.

An example

Assume the same determination and Shewhart chart as in Appendix B. Compare two data sets of results for the control standard given in Table I1. In both cases, the results found are within the 95%-confidence interval (i.e. between 90 and 110 g/kg), which means that no further action is required according to the criteria as described in Appendix B.

The CUSUM value is calculated as the difference between the result minus the average plus the previous CUSUM value. In the case of situation 1, the first CUSUM value is -5 (i.e. 95 100) and the second is 0 (i.e. 105 100 +(-5)). The CUSUM estimation however reveals a remarkable difference between both situations. In situation 1, the CUSUM value varies around zero, which is a normal result if only random variation occurs. In the second situation, however, the value clearly increases with time with the cumulative negative effect very apparent. This is an indication that a possible drift is present in the test, leading in this case to a systematic underestimation in the results found.

Although this method is easy to perform, deciding when to act is more complicated. In the literature, a complicated statistical method, known as V-masks, is described, but this is less suitable for use under normal laboratory conditions. A more pragmatic approach is to set a maximum value for each test, based on its SD and practical experience.

TABLE I1
Example of the estimation of the CUSUM values

Situation 1		Situation 2	
Result (g/kg)	CUSUM	Result (g/kg)	CUSUM
95	-5	99	-1
105	0	97	-4
99	-1	93	-11
97	-4	101	-10
107	3	94	-16
93	-4	99	-17
105	1	97	-20

FAO ANIMAL PRODUCTION AND HEALTH GUIDELINES

1. Collection of entomological baseline data for tsetse area-wide integrated pest management programmes, 2009 (E)
2. Preparation of national strategies and action plans for animal genetic resources, 2009 (E, F, S, R, C)
3. Breeding strategies for sustainable management of animal genetic resources, 2010 (E, F, S, R, Ar, C**)
4. A value chain approach to animal diseases risk management – Technical foundations and practical framework for field application, 2011 (E)
5. Guidelines for the preparation of livestock sector reviews, 2011 (E)
6. Developing the institutional framework for the management of animal genetic resources, 2011 (E, F, S)
7. Surveying and monitoring of animal genetic resources, 2011 (E, F, S)
8. Guide to good dairy farming practice, 2011 (E, F, S, R, Ar, Pt)
9. Molecular genetic characterization of animal genetic resources, 2011 (E)
10. Designing and implementing livestock value chain studies, 2012 (E)
11. Phenotypic characterization of animal genetic resources, 2012 (E, F)
12. Cryoconservation of animal genetic resources, 2012 (E)
13. Handbook on regulatory frameworks for the control and eradication of hpai and other transboundary animal diseases – A guide to reviewing and developing the necessary policy, institutional and legal frameworks, 2013 (E)
14. *In vivo* conservation of animal genetic resources, 2013 (E)
15. The feed analysis laboratory: establishment and quality control, 2013 (E**)

Availability: November 2013

Ar	– Arabic	Multil	–	Multilingual
C	– Chinese	*		Out of print
E	– English	**		In preparation
F	– French	e		E-publication
Pt	– Portuguese			
R	– Russian			
S	– Spanish			

The *FAO Animal Production and Health Guidelines* are available through the authorized FAO Sales Agents or directly from Sales and Marketing Group, FAO, Viale delle Terme di Caracalla, 00153 Rome, Italy.

Find more publications at
http://www.fao.org/ag/againfo/resources/en/publications.html